ADVANCED SEMICONDUCTOR DEVICES

Proceedings of the 2006 Lester Eastman Conference

SELECTED TOPICS IN ELECTRONICS AND SYSTEMS

Editor-in-Chief: **M. S. Shur**

Selected Topics in Electronics and Systems – Vol. 45

ADVANCED SEMICONDUCTOR DEVICES

Proceedings of the 2006 Lester Eastman Conference

Cornell, Ithaca, NY, USA   26 August 2006

Editors

Michael S. Shur

Rensselaer Polytechnic Institute, USA

Paul Maki

Office of Naval Research, USA

James Kolodzey

University of Delaware, USA

World Scientific

NEW JERSEY · LONDON · SINGAPORE · BEIJING · SHANGHAI · HONG KONG · TAIPEI · CHENNAI

Published by

World Scientific Publishing Co. Pte. Ltd.

5 Toh Tuck Link, Singapore 596224

USA office: 27 Warren Street, Suite 401-402, Hackensack, NJ 07601

UK office: 57 Shelton Street, Covent Garden, London WC2H 9HE

British Library Cataloguing-in-Publication Data

A catalogue record for this book is available from the British Library.

ADVANCED SEMICONDUCTOR DEVICES
Proceedings of the 2006 Lester Eastman Conference

ISBN-13 978-981-270-858-8
ISBN-10 981-270-858-8

Editor: Tjan Kwang Wei

Printed in Singapore.

DEDICATED TO MEMORY OF NAVAN PARTHASARATHY

PREFACE

This volume contains Proceedings of the 2006 biennial Lester Eastman Conference (LEC), which was held on the Cornell University campus on August 2-4, 2006. Originally, the conference was known as the IEEE/Cornell University Conference on High Performance Devices. It was renamed Lester Eastman Conference (LEC) to honor Prof. Lester Eastman, a renowned device pioneer and leader, in 2002 and held at the University of Delaware. The next LEC was held at the RPI campus in 2004 before coming back to Cornell in 2006.

Just after the conference, on the afternoon of Friday, August 4, 2006 a terrible tragedy cast a great sadness over the week's events. Mr. Navan Parthasarathy, a participant and a presenter at the LEC conference, had drowned in Fall Creek on Cornell Campus. He was a graduate student at the University of California, Santa Barbara. We dedicate the LEC-06 Proceeding to Mr. Navan Parthasarathy to honor his memory.

Professor Lester Eastman

UCSB Ph.D. student
Navan Parthasarathy
Passed away on August 4, 2006

The Proceedings cover five emerging and traditional areas of advanced device technology: wide band gap devices, terahertz and millimeter waves, silicon and silicon-germanium devices, nanoelectronics and ballistic devices, and photoluminescence and photocapacitance characterization of advanced photonic and electronic devices.

The papers by M. Sugimoto et al. entitled "Wide-bandgap semiconductor devices for automotive applications" and by B. Green et al. "A GaN HFET device technology for wireless infrastructure applications" define existing and future applications of the wide band gap electronic devices and, hence, set the stage for many outstanding papers describing physics, chemistry, device design, fabrication, and modeling of these device.

Papers on terahertz and millimeter include papers describing terahertz emission and sensing (from the Delaware group), the paper from Marc Rodwell's group on new millimeter wave phase array architecture, and the paper on millimeter wave heterostructure diode.

Papers dealing with Si and SiGe technology cover new device designs, fast trapping devices, and SiGeC/Si IR photonics. It is interesting to compare the paper on SiGeC/Si IR photonic with Stiff-Roberts' paper on hybrid nanomaterials for IR photo detection.

A thought provoking paper from Lester Eastman's and Brian Ridley's group discusses ballistic electron acceleration and negative differential conductivity devices.

M. Wraback et al. describe how dependent time-resolved photoluminescence helps understand the physics of ultraviolet AlN/GaN/InN based emitters.

All in all, these proceedings will bring the reader to the forefront of advanced device technology.

The Editors would like to thank the authors, anonymous reviewers who were also the key contributors to the success of these proceeding.

The conference was under the technical sponsorship of the Institute of Electrical and Electronic Engineering (IEEE). We are grateful to the Air Force Office of Scientific Research (AFOSR), the Defense Advanced Projects Agency (DARPA) and the Office of Naval Research (ONR), The Northrop Grumman Corporation, and the Cornell University College of Engineering for their support of the Lester Eastman Conference 2006.

EDITORS

Michael S. Shur (shurm@rpi.edu)
Paul Maki (Paul_Maki@onr.navy.mil)
James Kolodzey (kolodzey@ee.udel.edu)

CONTENTS

Section I.
Wide Bandgap Devices

International Journal of High Speed Electronics and Systems
Vol. 17, No. 1 (2007) 3–9
© World Scientific Publishing Company

WIDE-BANDGAP SEMICONDUCTOR DEVICES
FOR AUTOMOTIVE APPLICATIONS

M. Sugimoto

TOYOTA MOTOR CORP.
543, Kirigahora, Nishihirose-cho, Toyota, Aichi, 470-0309, JAPAN
E-mail : sugimoto@masahiro.tec.toyota.co.jp

H. Ueda, T. Uesugi, and T. Kachi

TOYOTA CENTRAL R&D LABS., INC.
Yokomichi, Nagakute, Aichi-gun, Aichi, 480-1192, JAPAN

In this paper, we discuss requirements of power devices for automotive applications, especially hybrid vehicles and the development of GaN power devices at Toyota. We fabricated AlGaN/GaN HEMTs and measured their characteristics. The maximum breakdown voltage was over 600V. The drain current with a gate width of 31mm was over 8A. A thermograph image of the HEMT under high current operation shows the AlGaN/GaN HEMT operated at more than 300°C. And we confirmed the operation of a vertical GaN device. All the results of the GaN HEMTs are really promising to realize high performance and small size inverters for future automobiles.

Keywords: GaN, HEMT, HV, inverter, normally-off, vertical device

1. Introduction

Development and improvement of hybrid vehicles (HVs), electric vehicles (EVs) and fuel cell hybrid vehicles (FCHVs) are now widely recognized as one of the solutions of the CO_2 problem of the earth and the exhaust gas problem of urban areas. These vehicles need high electric power inverters with high energy conversion efficiency. Si Insulated Gate Bipolar Transistors (Si-IGBTs) are widely used in the inverters, but these devices have a limitation of performance due to their material properties. Devices with higher performance have been strongly required for future vehicles. In this paper, we discuss requirements of automotive applications of wide-bandgap semiconductor devices, especially hybrid vehicles, and our development of GaN power devices.

Figure 1 shows the road map of the power density [1]. The power density has been increasing year after year. The inverters installed in Toyota's hybrid vehicles, such as Prius and RX400h have 3 times higher power densities than other applications as shown in the figure. However, HV and FCHV systems of the next generation will require much higher power densities with lower energy loss, smaller size and lower cost. It is difficult to realize such higher power density systems with Si-IGBTs, because of their material properties. Therefore, we should develop novel power devices made of new materials for these systems. Theoretical performance of several semiconductor materials are shown in Table 1, where wide-bandgap semiconductors clearly have advantages compared with Si. This is the big motivation for us to develop novel switching devices made of wide-bandgap semiconductors such as SiC and GaN.

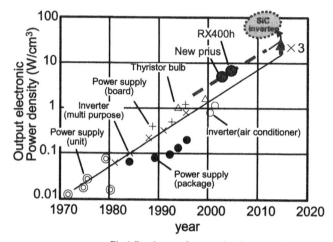

Fig.1 Road map of power density

Table 1. Normalized figures of merit of various semiconductors

	Si	GaAs	4H-SiC	GaN
JFM	1	11	410	790
KFM	1	0.45	5.1	1.8
BFM	1	28	290	910
BHFM	1	16	34	100

JFM   : Johnson's figure of merit for high frequency devices $=(E_b V_s/2\pi)^2$
KFM : Keyes's figure of merit considering thermal limitation $=\kappa(E_b V_s/2\pi\varepsilon)^{1/2}$
BFM : Baliga's figure of merit for power switching $=\varepsilon m E_g^3$
BHFM: Baliga's figure of merit for high frequency power switching$=\mu E_b^2$

2.   Requirements of Automotive Applications for Power Device in the Future

To realize the next generation hybrid systems, the following performance will be required for power devices in the future.

2.1. Normally-off operation
Most effort on GaN based devices has been directed toward normally-on ones, and only a few normally-off ones have been reported. In normally-off devices, the drain current does

not flow at the gate voltage of 0V. Si-IGBTs used in the present inverters are normally-off operation devices. Likewise, a normally-off operation device is required for future automobiles in order to simplify the inverter circuit and make effective use of design techniques and mounting technologies for the inverters.

2.2. High breakdown voltage

Figure 2 shows the relationship between the motor power of Toyota's HVs and power source voltages of these systems. The first generation Prius was on the market in 1997. Through the inverter of this hybrid system, the battery voltage of 277V was directly connected to the motor. The new Prius has been on the market since 2003. In its HV system, the battery voltage is once raised to a power source voltage by a voltage booster and then supplied to the motor through the inverter. The raised voltage can take values from 202V of the battery up to a maximum of 500V [2]. The new Toyota's HVs need high motor power with high power source voltage as shown in Fig. 2. The breakdown voltage of devices used in these inverters is about 1.1kV [3]. The breakdown voltage of devices used in the inverters will probably become higher in the future due to protection against surge voltage and so on. On the other hand, A upper limit of the breakdown voltage of them may depend on the withstand voltage of condensers, discharge inception voltages between phases of the motors, etc.

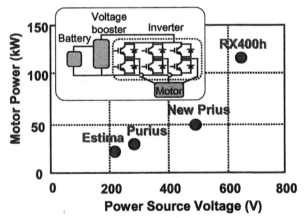

Fig.2 Power source voltage vs. motor power

2.3. Low on-resistance and high current density

In order to increase the energy conversion efficiency of inverter systems, it is necessary to decrease the on-resistance of the devices. Moreover, it is demanded for miniaturization of inverters to increase the current capacity up to several hundreds A/cm^2.

2.4. High temperature operation

Si devices stop working over 150°C, because a power loss increases due to an increase of a leakage current in the off state under high temperature envirnoments. Whereas, GaN devices can probably work over 200°C, because the energy bandgap of GaN is wider than that of Si. The present hybrid systems have two cooling systems, one is for the engine and the other is for the inverter, and the temperature of coolant water for the inverter is lower than that for the engine. If GaN devices work over 200°C, the cooling system of hybrid vehicles will be simplified and its cost will be reduced.

2.5. Vertical operation

Up to date, most of efforts on development of GaN based devices have been directed toward lateral operation ones. However, vertical operation devices are demanded for automotive applications. First, vertical devices will meet the requirement of the above 3). Secondly, design techniques and mounting technologies developed for Si IGBT will be utilized effectively.

3.   Our Development of GaN Power Devices

3.1. Normally-off operation

Studies of GaN based device have been mainly focused on the normally-on operation devices for microwave power applications, for example, the base station. Recently, a few normally-off operation GaN devices for power electronics applications have been reported, and the threshold voltages (Vth) of these devices are around +0.3V [4] [5]. However, no device with a high drain current and a very low on-resistance has been realized yet, because most of these devices are Schottky gate type. In the Schottky type devices, a current begins to flow drastically from the gate to the source at the gate voltage of around +2V. Therefore, it is impossible to control high current by the Schottky gate electrode. On the other hand, in case of insulated gate type devices, the higher voltage than around +2V can be applied to the gate electrodes. In order to realize the normally-off devices with the high current operation and the low on-resistance, the key issue is formation of gate insulator films with good quality. We investigated SiO_2 films formed by Low Pressure Chemical Vapor Deposition (LP-CVD) at a high temperature [6]. Results of the interface-state density calculated by the Terman method are shown in Fig. 3. The curve of the interface state density had two peaks at about -0.4eV and -0.7eV from

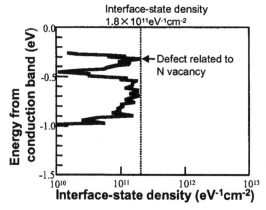

Fig. 3 SiO_2/GaN interface-state density

the conduction band, and both of the interface state densities were about $2 \times 10^{11}/cm^2eV$. We are developing normally-off devices using SiO_2 films. We will show some of their characteristics next.

3.2. High breakdown voltage

A GaN-HEMT with break down voltage of 1.3kV was reported in recent years [7]. We fabricated the insulated gate HEMTs, and measured the breakdown characteristics. Figure 4 shows the result of breakdown characteristic at the gate bias of -30V. The gate-drain length and the gate width of this device are 20μm and 440μm, respectively. The breakdown voltage was over 600V.

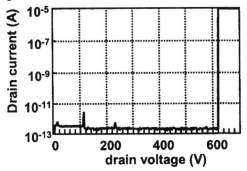

Fig. 4 Breakdown characteristics

3.3. Low on-resistance and high current operation

To examine a possibility of GaN power devices, we fabricated a large size GaN power HEMT with a gate width of 31 mm (Fig. 5), and verified a high current operation [6]. The drain current reached over 8A under pulse measurement (Fig.6). The specific on-resistance was about $5m\Omega$-cm^2.

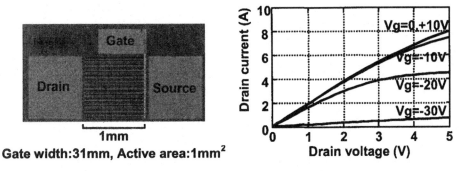

Gate width:31mm, Active area:1mm²

Fig. 5 Photograph of fabricated GaN-HEMT

Fig. 6 I_D-V_D characteristics under pulse condition

3.4. High temperature operation

An observation on the device under high current operation with thermograph (Fig.7) showed the GaN HEMT operated at more than 300°C [8]. The gate width and the chip size of this device were 31mm and 2x4mm^2, respectively. The active area was about 1x1mm^2. The bias condition was V_D=2.6V and I_D=1.3A under DC condition. These results indicate that GaN power devices are promising for future automotive systems operated in high temperature circumstance.

3.5. Vertical operation

In order to decrease the on resistance, we also fabricated a vertical operation device on a free-standing n-type GaN substrate[9]. Figure 8 shows the current-voltage characteristics. A drain current didn't rise at a voltage of 0V because of poor characteristics of ohmic contacts. But by using a free-standing GaN substrate, vertical current between source and drain was really well controlled by the gate bias. A device with a gate length of 4μm and a gate width of 300μm showed a leakage current in off state was less than 10^{-9}A.

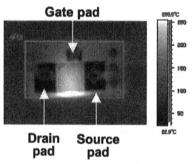

Fig. 7 Thermograph image under high current operation

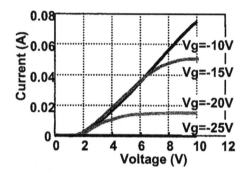

Fig. 8 I-V characteristics of vertical operation device

4. Conclusions

We explained the inverter in the Toyota's hybrid systems and requirements of future hybrid systems for power devices. And we find that GaN power devices are excellent candidates for future automotive applications.

We fabricated AlGaN/GaN HEMTs and measured their characteristics. The maximum of the breakdown voltage was over 600V. The drain current was over 8A at a device with a gate width of 31mm. All the results of our GaN technology are really promising to realize high performance small size inverter for future automotive applications.

On the other hand, many issues which should be solved also become clear, for example, an improvement of the breakdown voltage, an achievement of good balance between a reduction of the specific on-resistance and the normally-off operation and a realization of vertical operation devices.

Other important issues we don't refer are an evaluation and a securing of long-term stability and a development of free-standing GaN substrate with lower defect density, larger size and lower cost.

5. Acknowledgements

The authors would like to thank O. Ishiguro for epitaxial growth and Dr. M. Kodama and E. Hayashi for SiO_2 film deposition and N. Soejima, Dr. M. Kanechika and T. Terada for device fabrication and measurements. Additionally, the authors gratefully acknowledge the many suggestions and support by Dr. H. Tadano.

References

1. H. Ohashi et al., "Power Electronics Innovation with Next Generation Advanced Power Devices", IEICE TRANS. COMMUN., VOL.E87-B, NO.12, pp-3422-3429, 2004
2. H. Kawahashi, "A New-Generation Hybrid Electric Vehicle and Its Supporting Power Semiconductor Devices", ISPSD 2004 Tech. Digest, pp23-29, 2004
3. TOYOTA Technical Review vol.54, No.1, 2005
4. V. Kumar et al., "High transconductance enhancement-mode AlGaN/GaN HEMTs on SiC substrate", Electronics Letters vol.39, No.24, pp1758-1759, 2003
5. N. Ikeda et al., "Normally-off Operation Power AlGaN/GaN HFFT", ISPSD 2004 Tech. Digest, pp369-372, 2004
6. M. Sugimoto et al., "A Study of MIS-AlGaN/GaN HEMTs with SiO_2 Films as Gate Insulator", ISPSD 2005 Tech. Digest, pp307-310, 2005
7. N. Q. Zhang et al., "Effects of surface traps on breakdown voltage and switching speed of GaN power switching HEMTs", IEDM 2001 Tech. Digest, pp587-590, 2003
8. H. Ueda et al., "High Current Operation of a GaN Power HEMTs", ISPSD 2005 Tech. Digest, pp311-314, 2005
9. T. Kachi et al., "Vertical device operation of AlGaN/GaN heterostructure field effect transistor fabricated on GaN substrates", ICNS 2005 Th-P-139, 2005

International Journal of High Speed Electronics and Systems
Vol. 17, No. 1 (2007) 11–14
© World Scientific Publishing Company

A GAN ON SIC HFET DEVICE TECHNOLOGY FOR WIRELESS INFRASTRUCTURE APPLICATIONS

B. Green, H. Henry, K. Moore, J. Abdou, R. Lawrence, F. Clayton, M. Miller, J. Crowder, E. Mares, O. Hartin, C. Liu, C. Weitzel

Freescale Semiconductor, Inc

2100 E. Elliot Rd. MD EL720 , Tempe, Arizona 85284
Bruce.M.Green@freescale.com

This paper presents Freescale's baseline GaN device technology for wireless infrastructure applications. At 48 V drain bias and 2.1 GHz operating frequency 10-11 W/mm, 62-67% power-added efficiency (PAE) is realized on 0.3 mm devices and 74 W (5.9 W/mm), 55% PAE is demonstrated for 12.6 mm devices. A simple thermal model shows that a more than twofold increase in channel temperature is responsible for limiting the CW power density on the 12.6 mm compared to 0.3 mm devices. The addition of through wafer source vias to improve gain and tuning the device in a fixture optimized for efficiency yield an output power of 57W (4.7 W/mm), PAE of 66%, and a calculated channel temperature of approximately 137˚C at a 28 V bias. *Keywords*: GaN, SiC, Transistors.

1. Introduction

The properties of the AlGaN/GaN material system have enabled demonstration of astonishing microwave power densities with clear system performance advantages for wireless infrastructure applications[1]. Despite the tremendous advantages afforded by these high power densities, care must be taken in the design and application of the technology to maintain channel temperatures consistent with reliability and packaging requirements that often demand channel temperatures of less than 200˚C. This paper describes Freescale's baseline GaN technology, characteristics of small devices followed by the characteristics of high power amplifiers. The results are analyzed using a simple thermal model showing an estimated channel temperature that is approximately 2.5 X for 12.6 mm devices as compared to 0.3 mm devices at 48 V bias. It is then shown that by improving the gain and power-added efficiency (PAE) using a through wafer via process and then testing the devices in a 50Ω fixture optimized for efficiency that the peak channel temperature is reduced to a level acceptable for low cost packaging and reliability.

2. Freescale GaN HFET Technology and Small Device Performance

The technology employs an un-doped AlGaN/GaN HFET structure grown on 3" SiC semi-insulating substrates with a sheet resistance of approximately 380 ohms/sq. Device processing includes ECR mesa etch isolation, Ti/Al/Mo/Au ohmic contacts, SiN passivation[2], Ni-Au gates, and plated Au air-bridges. After front-side processing, the SiC substrates are thinned and back-metallized using a Au plating process.

The power density and efficiency of any transistor is driven by the RF voltage-current characteristics. Pulsed I-V curves are an indicator of RF power and efficiency since the short pulses tend to "freeze out" carriers trapped by bulk and surface traps. As seen in Figure 1(a), very good agreement is seen between the pulsed and DC behavior for Freescale's GaN HFET process. In addition, the negligible DC-RF current slump seen for this technology produces 2.1 GHz, 48V saturation characteristics free of pre-mature gain compression as seen in Figure 1(b).

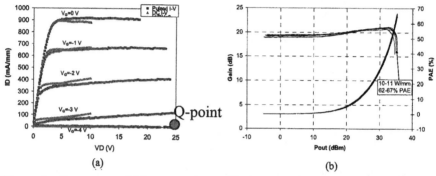

Figure 1: Summary of (a) comparison of DC and Pulsed I-V characteristics (pulse width 500 nS, V_{DQ}=25 V, V_{GQ}=-5V) and (b) 2.1 GHz, 48 V power sweeps at for 2×150 μm GaN HFET's (I_{DQ}=140 mA/mm).

3. High Power Amplifier Performance and Thermal Analysis

After singulating into die, processed large periphery devices with back-metal were bonded using Au-Sn pre-forms into industry-standard NI-360 metal-ceramic packages for large signal testing. Bondwires were used to ground the sources of the device to the package. Figure 2(a) shows 2.14 GHz CW drive up characteristics for a 12.6 mm device at a drain bias of 28V. As can be seen from the data, a saturated output power of 63 W (5 W/mm) is achieved at a 28 V drain bias while an output power of 74 W (5.9 W/mm) is achieved at a drain bias voltage of 48 V as shown in Figure 2(b).

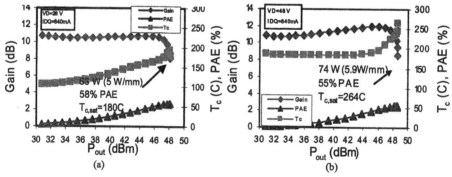

Figure 2: Saturation characteristics of 12.6 mm non-source via devices at 2.14 GHz biased at (a) 28 V and (b) 48 V.

As is the case for the small periphery device data in the previous section, the gain saturation characteristics are well-behaved with very little pre-mature gain compression evident. However, compared to the small periphery device data shown in Fig. 1(b), the microwave power density is just over half at 5.9 W/mm at 48 V drain bias for the large device.

3.1. *Thermal Analysis*

To evaluate the effect of channel temperature, the plots of large signal saturation characteristics of Fig. 2 also show the estimated channel temperature based on a simple expression for the channel temperature given by

$$T_c = P_{DISS}\left[P_{DISS}\left(\frac{R_{TH2} - R_{TH1}}{P_{DISS2} - P_{DISS1}}\right) + R_{TH1}\right] + T_{Stage} \tag{1}$$

where R_{TH1} and R_{TH2} are the thermal resistances evaluated using a calibrated thermal model at the power dissipation levels P_{DISS1} and P_{DISS2} respectively. In this case, R_{TH1} and R_{TH2} are evaluated at power dissipation levels of 3 W/mm and 5 W/mm respectively. The stage temperature is 27°C. In addition to accounting for the temperature rise of the channel with respect to the case, the thermal resistances also account for the rise in the case temperature with respect to the stage on which the device fixture is mounted. As seen from the data in the plots, the channel temperature is approximately 180°C for the case of 28 V drain bias and is approximately 264°C for the case of 48 V drain bias. Re-examination the small periphery device data of Fig. 1(b) using the thermal model of Eq. (1) with appropriately scaled parameters shows that the maximum channel temperature for 48 V operation is significantly lower with a value of only 102°C achieved at saturation as seen in Figure 3. Since the mobility of GaN has a $T^{1.8}$ dependence [3] the 2.5X higher channel temperature (i.e. 264 °C compared to 102°C) for the 12.6 mm device increases the portion of the on resistance of the device attributable to the device sheet resistance by a factor of approximately 5.5 thereby reducing the current and increasing the knee voltage.

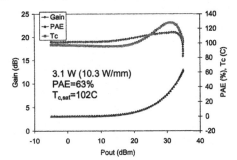

Figure 3: Saturation characteristics of 0.3 mm non-source via devices at 2.1 GHz showing the estimated channel temperature of the device.

The significantly higher channel temperature seen for the case of 48 V bias compared to the 28 V case explains the modest increase in power density in going from 28 V to 48 V drain bias. Given the fact that low-cost plastic packaging demands maximum channel temperatures of less than 200°C and that the actual application of these devices is typically at a 90°C case temperature, it is highly desirable to reduce the channel temperature. An expression for the dissipated power, P_{DISS}, given by

$$P_{DISS} = P_{DC}(1 - PAE) = V_{DS}I_{DS}\left[1 - \eta_D\left(1 - \frac{1}{G_T}\right)\right] \tag{2}$$

shows that P_{DISS} can be reduced by increasing the drain efficiency, η_D or transducer gain, G_T.

3.2. *Effect of Source Vias on Gain and Thermal Performance*

Additional devices were fabricated using a through wafer via process developed at Freescale. As seen from the 2.14 GHz loadpull data of Figure 4(a), the addition of a source via to the process provides a 2 dB gain improvement over devices without vias. The devices were then mounted in a fixture optimized for class A-B efficiency. As seen in Figure 4(b), an output power of 58 W and 66% PAE was achieved with a maximum estimated channel temperature of 137°C.

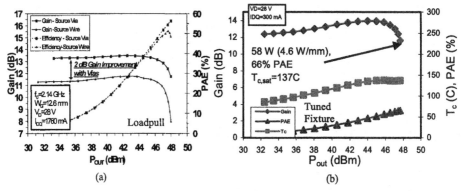

(a)   (b)

Figure 4: Large signal characteristics of 12.6 mm device biased at V_{DS}=28V (a) with and without source vias (loadpull result) and (b) in a tuned test fixture optimized for class A-B efficiency.

4. Conclusion

This paper has presented the results of Freescale's baseline GaN HFET process and identifies thermal effects as the limiting factor for scaling the power densities from 0.3 mm devices to 12.6 mm devices. Furthermore, the importance of improved PAE through device design and high efficiency power amplifier design for achieving acceptable channel temperature has been established.

References

1. Y. Wu, et al, "30 W/mm GaN HEMT's by Field Plate Optimization" *IEEE Elect. Dev. Lett.*, November 2004.
2. B. Green, et al, "Effect of Surface Passivation on the Microwave Characteristics of AlGaN/GaN HEMT's", *IEEE Elect. Dev. Lett.*, August 2000.
3. L.F. Eastman, Private Communication.

International Journal of High Speed Electronics and Systems
Vol. 17, No. 1 (2007) 15–18
© World Scientific Publishing Company

Drift velocity limitation in GaN HEMT channels

ARVYDAS MATULIONIS*

Semiconductor Physics Institute, A. Goštauto 11, Vilnius 01108
Lithuania

Additional friction due to Pauli constraint, channel self-heating, alloy scattering, and hot phonons is reconsidered.

Keywords: Two-dimensional channels, GaN, AlGaN, drift velocity; hot electrons; hot phonons.

1. Introduction

An undoped GaN high-electron-mobility transistor (HEMT) with a two-dimensional electron gas (2DEG) channel is a promising high-power active device [1]. Transistor unity-gain cut-off frequency exceeds 150 GHz for short-gate devices[2,3]. A linear dependence of the inverse cut-off frequency on the gate length yields an effective electron drift velocity. The extracted velocity ranges from $1 \cdot 10^7$ cm/s to $1.75 \cdot 10^7$ cm/s [1,2,3,4,5]. The drift velocity decreases as the electron density increases[5]. Several times higher values follow from time-of-flight experiments on GaN p-i-n diodes[6]. Thus, drifting electrons suffer additional friction in biased HEMT channels. The following effects can be of importance at a high electron density: (i) 2DEG degeneracy, (ii) channel self-heating at a high density of current and a high bias, (iii) alloy scattering caused by an overlap of electron envelope function with that of random alloy potential, (iv) accumulation of non-equilibrium optical phonons (termed hot phonons). Our goal is to discuss recent advances in understanding of electron drift velocity limitation in nominally undoped 2DEG channels for high-power GaN HEMTs.

2. Pauli constraint

Scattering events into occupied states are forbidden in a degenerate 2DEG channel. As a result, the electron drift velocity tends to increase. Monte Carlo simulation for a

*matulionis@pfi.lt

GaN 2DEG channel shows[7] that this effect prevails at low electric fields. Unlike this, when polar optical phonon scattering becomes the dominant one at a higher electric field, the Pauli constraint excludes predominantly small-angle scattering events of optical phonon emission, and the leading electrons suffer large-angle scattering: the drift velocity decreases when 2DEG degeneracy is taken into account[7].

3. Channel self-heating

Microwave noise technique is used to measure noise temperature during and after the voltage pulse. Under bias, a monotonous increase in the channel temperature is accompanied with a time-dependent decrease in the noise temperature[8]. When the bias is off, the measured noise temperature relaxation and its back extrapolation yields the channel temperature under the bias at the end of the voltage pulse. A typical relaxation time constant is 400 ns [8]. A negligible self-heating effect on electron drift velocity is observed at electric fields below 80 kV/cm for 150 ns voltage pulses[9]. In the range up to 300 kV/cm, the effect is weak for the pulses lasting several nanoseconds.

4. Alloy scattering

Consider a 2DEG channel located in a GaN layer of an AlGaN/GaN heterostructure. A substitution of an Al ion for a Ga ion introduces a local random potential; it acts as a scattering centre for drifting electrons. When hot electrons penetrate into the adjacent AlGaN layer, their contribution to drift velocity changes, and the associated noise appears[10]. The threshold field for this source of noise correlates with the heterojunction barrier height[11]. The noise due electron penetration is avoided when AlN spacer is used[9]. Remote alloy scattering is caused by penetration of the random potential into the 2DEG channel. A combination of a GaN spacer with an AlN spacer in a GaN-spacer HEMT[12] reduces the penetration of the random potential into the 2DEG channel. As a result, the electron drift velocity increases[11,12] (Fig. 1, stars and bullets).

5. Hot phonons

Hot electrons launch non-equilibrium longitudinal optical phonons (termed hot phonons). According to Monte Carlo simulation, hot phonons introduce additional friction for drifting electrons (Fig. 1, squares and triangles). The effect is weaker if the hot-phonon lifetime is shorter[14]. Time-resolved pump-probe Raman technique is the most direct technique for hot-phonon lifetime measurement in bulk semiconductors. A density-independent value (τ_{ph}=2.5 ps) at low electron densities ($\sim 10^{16}$cm^{-3}) is followed with lower values at higher densities until τ_{ph}=350 fs at $\sim 2 \cdot 10^{19}$ cm^{-3} is recorded for bulk GaN samples[16].

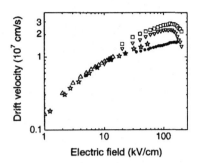

Fig. 1. Dependence of electron drift velocity on electric field for GaN 2DEG channels. Experiment: AlGaN/GaN (bullets)[13] and AlGaN/GaN/AlN/GaN (stars)[11]. Monte Carlo simulation: τ_{ph}=350 fs (up triangles)[14], τ_{ph}=300 fs (down triangles)[15], and hot phonons neglected (squares)[15].

The hot-phonon lifetime is not measured in a biased 2DEG channel through the time-resolved pump-probe Raman technique. Noise technique[14] has yielded τ_{ph}=350 fs for a 2DEG in AlGaN/GaN at effective electron density $\sim 1 \cdot 10^{19}$ cm^{-3}. For a similar AlGaN/GaN structure, a close value of 380 fs is obtained from time-resolved intersubband absorption experiment[17]. Thus, in the 2DEG channel located in GaN[14,17], the lifetime nearly coincides with that obtained for a bulk GaN sample[16] supposing that the electron densities are similar. No essential dependence of the lifetime on the electron temperature and the channel temperature is found[18].

Acknowledgments

The US Office of Naval Research (Grant N00014-03-1-0558 monitored by Dr. Colin E.C. Wood), the European Commission within the Network of Excellence SINANO (IST- 506844), and the Lithuanian National Foundation for Science and Education (Contract LVMSF C-33/2006) are gratefully acknowledged for support.

References

1. L. F. Eastman, V. Tilak, J. Smart, B. M. Green, E. M. Chumbes, R. Dimitrov, H. Kim, O. S. Ambacher, N. Weimann, T. Prunty, M. Murphy, W. J. Schaff, R. Shealy, "Undoped AlGaN/GaN HEMTs for microwave power amplification", *IEEE Trans. ED* **48** (2001) 479–485.
2. T. Palacios, A. Chakraborty, S. Heikman, S. Keller, S. P. DenBaars, and U. K. Mishra, "AlGaN/GaN high electron mobility transistors with InGaN back-barriers", *IEEE Electron Device Letters* **27** (2006) 13–15.
3. M. Higashiwaki, T. Matsui, and T. Mimura, "AlGaN/GaN MIS-HFETs with f_T of 163 GHz using cat-CVD SiN gate-insulating and passivation Layers", *IEEE Electron Device Letters* **27** (2006) 16–18.
4. T. Inoue, Y. Ando, H. Miyamoto, T. Nakayama, Y. Okamoto, K. Hataya, M. Kuzuhara, "30-GHz-band over 5-W power performance of short channel AlGaN/GaN heterojunction FETs", *IEEE TMTT* **53** (2005) 74–80.

5. C. H. Oxley and M. J. Uren, A. Coates, and D. G. Hayes, "On the temperature and carrier density dependence of electron saturation velocity in an AlGaN/GaN HEMT", *IEEE Trans. Electron Devices* **53** (2006) 565–567.

6. M. Wraback, H. Shen, J. C. Carrano, C. J. Collins, J. C. Campbell, R. D. Dupuis, M. J. Schurman, I. T. Ferguson, "Time-resolved electroabsorption measurement of the transient electron velocity overshoot in GaN", *Appl. Phys. Lett.* **79** (2001) 1303–1305.

7. M. Ramonas, A. Matulionis, L. Rota, "Monte Carlo simulation of hot-phonon and degeneracy effects in the AlGaN/GaN two-dimensional electron gas channel", *Semicond. Sci. Technol.* **18** (2003) 118–123.

8. L. Ardaravičius, J. Liberis, A. Matulionis, L. F. Eastman, J. R. Shealy, and A. Vertiatchikh, " Self-heating and microwave noise in AlGaN/GaN" *phys. stat. sol. (a)* **201** (2004) 203–206.

9. M. Ramonas, A. Matulionis, J. Liberis, L. Eastman, X. Chen, and Y.-J. Sun, " Hot-phonon effect on power dissipation in a biased AlGaN/AlN/GaN channel", *Phys. Rev. B* **71** (2005) 075324-1–8.

10. A. Matulionis and J. Liberis, "Microwave noise in AlGaN/GaN channels", *IEE Proc.-Circuits Devices Syst.* **151** (2004) 148–154.

11. A. Matulionis, J. Liberis, O. Kiprijanovic, T. Palacios, A. Chakraborty, S. Keller, and U. K. Mishra, "Effect of alloy scattering on electron drift velocity in GaN HEMTs", in *Abstract book WOCSDICE-2006*, Ed. Jan Stake, Chalmers University of Technology, ISSN 1652-0769, 2006, pp. 165–166.

12. T. Palacios, L. Shen, S. Keller, A. Chakraborty, S. Heikman, D. Buttari, S. P. DenBaars, and U. K. Mishra, "Demonstration of a GaN-spacer high electron mobility transistor with low alloy scattering", *phys. stat. sol. (a)* **202** (2005) 837–840.

13. L. Ardaravičius, M. Ramonas, O. Kiprijanovic, J. Liberis, A. Matulionis, L. F. Eastman, J. R. Shealy, X. Chen, and Y. J. Sun, "Comparative analysis of hot-electron transport in AlGaN/GaN and AlGaN/AlN/GaN channels", *phys. stat. sol. (a)* **202** (2005) 808–811.

14. A. Matulionis, J. Liberis, I. Matulionienė, M. Ramonas, L. F. Eastman, J. R. Shealy, V. Tilak, and A. Vertiatchikh, "Hot-phonon temperature and lifetime in a biased $Al_xGa_{1-x}N$/GaN channel estimated from noise analysis", *Phys. Rev. B* **68** (2003) 035338-1–7.

15. J. M. Barker, D. K. Ferry, S. M. Goodnick, D. D. Koleske, A. Allerman, and R. J. Shul, "High field transport in GaN/AlGaN heterostructures", *J. Vac. Sci. Technol. B* **22** (2004) 2045–2050.

16. K. T. Tsen, J. G. Kiang, D. K. Ferry, and H. Morkoc, "Subpicosecond time-resolved Raman studies of LO phonons in GaN: Dependence on injected carrier density", *Appl. Phys. Lett.*(2006) 112111-1–3.

17. Z. Wang, K. Reimann, M. Woerner, T. Elsaesser, D. Hofstetter, J. Hwang, W. J. Schaff, and L. F. Eastman, "Optical Phonon Sidebands of Electronic Intersubband Absorption in Strongly Polar Semiconductor Heterostructures", *Phys. Rev. Lett.* **94** (2005) 037403-1–4.

18. A. Matulionis, J. Liberis, L. Ardaravičius, L. F. Eastman, J. R. Shealy, and A. Vertiatchikh, "Hot-phonon lifetime in AlGaN/GaN at high lattice temperatures", *Semicond. Sci. and Technol.* **19** (2004) S421–S423.

International Journal of High Speed Electronics and Systems
Vol. 17, No. 1 (2007) 19–23
© World Scientific Publishing Company

SIMULATIONS OF FIELD-PLATED AND RECESSED GATE GALLIUM NITRIDE-BASED HETEROJUNCTION FIELD-EFFECT TRANSISTORS

Valentin O. Turin and Michael S. Shur

Department of Electrical, Computer and Systems Engineering, Rensselaer Polytechnic Institute
Troy, New York 12180, USA

Dmitry B. Veksler

Department of Physics, Applied Physics and Astronomy, Rensselaer Polytechnic Institute
Troy, New York 12180, USA

We report on two-dimensional isothermal simulations of recessed gate and field-plated AlGaN-GaN HFETs with submicron gates. The optimization of the recessed gate shape allows us to reduce the electric field at the drain-side gate edge by approximately 30%. Our simulations reveal a dramatic increase of the effective gate length with increasing drain-to-source bias with a commensurate decrease of the cutoff frequency (up to 40% decrease for 50V). To improve the cutoff frequency for the high drain-to-source bias, we suggest using the second field plate connected to the drain with a small gap between the two field plates. In this design, the electric field in the gap between the gate and the drain field plate is higher leading to a significant reduction of the effective gate length and, as a consequence, to an increase in the cutoff frequency at high drain-to-source biases (compared to the conventional design).

Keywords: Gallium Nitride; heterojunction field-effect transistor; cutoff frequency.

1. Introduction

Over the past several years, GaN-based field-effect transistors have demonstrated potential for high-voltage and high microwave power applications. However, their reliability still limits their applications in today's electronic systems. The field-plated [1] and recessed gate [2] AlGaN-GaN heterojunction field-effect transistors (HFETs) show improved performance due to the electric field reduction in the device channel and surface modification. Further improvement can be obtained using an additional drain field plate (FP) [3,4] and gate with facet on its drain side (i.e. gate edge engineering). We report on two-dimensional simulations of field-plated and recessed gate AlGaN-GaN HFETs with submicron gates. We ran isothermal simulations in the frame of drift-diffusion model and compared different gate and FP designs. The simulations were performed using commercially available Sentaurus Device software from Synopsys, Inc.

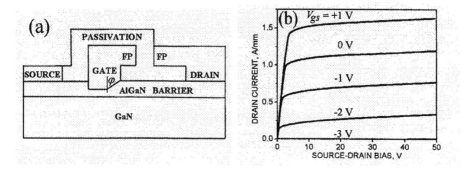

Fig. 1. (a) Schematic of simulated AlGaN-GaN HFET with recessed gate and with gate and drain field plates. The angle of a facet at drain side of the gate is φ. (b) - Simulated output characteristics.

Our simulations provide insight into the device physics and allow for the design optimization. The proposed recessed gate design with two field plates – gate and drain – and with a facet shows a reduced field at the drain-side gate edge, a fairly uniform field distribution in the channel and an improved cutoff frequency dependence on the drain-to-source bias.

2. Simulated Device

A schematic picture of simulated AlGaN-GaN HFET is shown on Fig. 1 (a). The source-drain separation is 3.75 μm. The source-gate separation is 0.9 μm. The gate length is 0.15 μm. The thickness of the undoped GaN layer is 2 μm. All simulations are performed for the 10^{18} cm^{-3} doping concentration in the AlGaN barrier layer with 18 nm thickness. For simulations of HFET with recessed gate, the recess depth of 18 nm was selected (the 36 nm total thickness of AlGaN layer was close to the value reported in [2]). We assume the 0.28 mole fraction of Al in AlGaN. The transistor was passivated with Si$_3$N$_4$ with 100 nm thickness. We simulated gate field plates with the length from 0.1 to 0.4 μm. We found that 0.15 μm gate FP gives a fairly uniform field distribution in the channel and we used this length in the simulations presented in this paper. Details about the boundary conditions used are described, for example, in [5] or in the Sentaurus Device manual [6]. We assumed the Schottky barrier potential height to be 1.5 eV, the electron mobility of 1200 cm^2/Vs, and the electron saturation velocity, v_{sat}, of 2 x 10^7 cm/s. Source and drain contacts resistances were 0.3 Ω·mm each. We assumed the fixed charge density at the AlGaN-GaN interface 1.1 x 10^{13} cm^{-2} to account for spontaneous and piezoelectric polarizations. We also introduced acceptor type traps with densities of 10^{17} cm^{-3} in GaN and 5 x 10^{16} cm^{-3} in AlGaN with the electron and hole capture cross sections of 10^{-15} cm^2.

3. Numerical Simulations

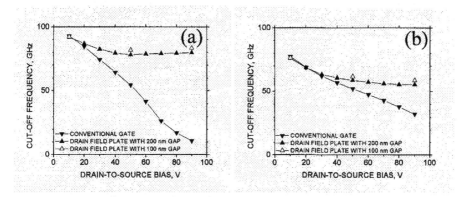

Fig. 2. Cutoff frequency dependence on drain-to-source bias for AlGaN-GaN HFET (a) - for conventional gate without drain FP (lower curve) and with drain FP (upper curve) and (b) - for recessed field-plated gate with 70^0 facet without drain FP (lower curve) and with drain FP (upper curve).

The simulated output characteristics of AlGaN-GaN HFET are presented in Fig. 1b. In Fig. 2a, the lower curve shows the simulated cutoff frequency dependence on drain-to-source voltage, V_{DS}, for the conventional gate. A decrease of the cutoff frequency with rising drain-to-source bias was approximately 40% for V_{DS}=50V. We estimated the effective gate length, $L_{g\ eff}$, from the simulated cutoff frequency f_t (see Table I):

$$L_{g\ eff} = \frac{v_{sat}}{2\pi f_t}$$

Although FPs mostly serve to increase the breakdown voltage, they can be used to influence electron velocity distribution in channel that, in turn, might improve the cutoff frequency [7]. For this propose, we use the FP connected to the drain with small gap between the drain FP and the gate FP. In this case, the high field is mostly localized in the gap between the gate and the second FP with the field distribution similar to that of a dipole formed by two charged wires (compare Fig. 3a and 3b). The result is a much smaller increase of the effective gate length (compare Fig. 4a and 4b). For such a design, the cutoff frequency dependence on the drain bias was significantly improved (with only 15% decrease for the 50 V drain-to-source bias for a 0.2 μm gap between the gate edge and the drain FP).

However, this design increases the maximum value of the electric field in the device channel, which might negatively affect the device reliability (see Fig. 3c). To counteract field increase, a more advanced design with the recessed field-plated gate with a facet at the drain side can be used. Our simulations show that the optimal facet angle φ is close to 70^0. Such a recessed gate design with a FP with 0.15 μm length results in the lower field at the drain-side gate edge and a fairly uniform field distribution in the high field region (see Fig. 3d and 3f). In this design, the maximum field in the channel is about 30% smaller compared to the conventional gate (compare solid lines in Fig. 3c and Fig. 3f).

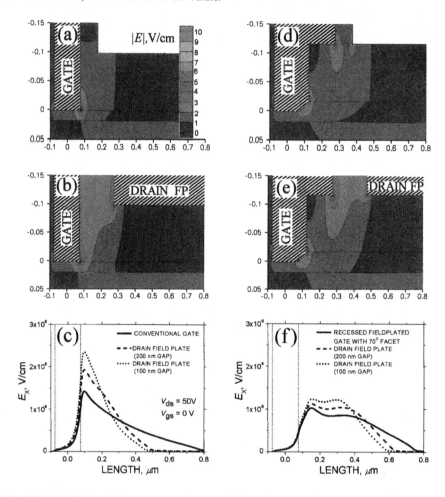

Fig. 3. Field distribution (V/cm) at gate edge for the case of: (a) non-recessed gate; (b) non-recessed gate with drain FP; (d) recessed field-plated gate with 70^0 facet; (e) recessed field-plated gate with 70^0 facet with drain FP. Gap between drain field plate and gate metal edge is 0.2 μm. Drain-to-source bias is 50V and gate-to-source bias is zero. (c) x – component of the channel electric field for non-recessed gate design with drain FP (dashed and dotted lines) and without drain FP (solid line); (f) the same dependence for field-plated recessed gate with 70^0 facet with drain FP (dashed and doted lines) and without the drain FP (solid line).

For the recessed gate, adding the drain FP also led to an improved cutoff frequency dependence on drain-to-source bias (see the upper curve in Fig. 2b) and only to a moderate increase of the maximum electric field at the drain side of the gate (see Fig. 3e and dashed and dotted lines in Fig. 3f). (In contrast, simulation of power GaN-based HFETs (with a much larger separation between the gate and drain FPs) reported in [3,4] predicted the peak electric field reduction, since the separation between the gate and drain field plates in that case was much larger than the extension of the high field region into the gate-to-drain spacing).

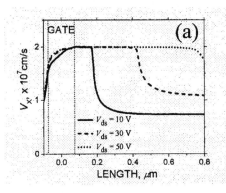 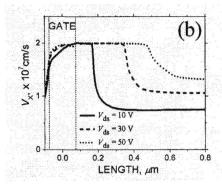

Fig. 4. Distribution of electron drift-velocity in channel (1 nm under AlGaN-GaN interface) is shown for different drain-to-source biases for (a) conventional gate without drain FP and (b) with drain FP (gap between gate and drain FP is 0.2 μm).

In conclusion, to improve the cutoff frequency for the high drain-to-source bias, we proposed using the drain FP with a relatively small gap between the gate and drain FPs. For such design, the high field is localized in this gap resulting in the noticeable decrease of the effective gate length at high drain biases and in a commensurate increase of the cutoff frequency.

This work was supported by the National Science Foundation under MURI "EMMA" and under IGERT (Grant No. 0333314). Any opinions, findings, and conclusions or recommendations expressed in this material are those of the authors and do not necessarily reflect the views of the National Science Foundation.

Table 1. f_T and the effective gate length for conventional gate without and with drain FP (with 0.2 μm gap between gate and drain FP).

V_{ds}, V	f_t, GHz	$L_{g\,eff}$, μm	f_t, GHz (Drain FP)	$L_{g\,eff}$, μm (Drain FP)
10	93	0.34	92	0.35
30	74	0.43	83	0.39
50	54	0.59	78	0.41
90	11	2.1	80	0.40

References

1. J. Li, S.J. Cai, G.Z Pan, Y.L. Chen, C.P. Wen, and K.L. Wang, *Electron Lett.* **37**, 196-197 (2001).
2. Y. Okamoto, Y. Ando, K. Hataya, T. Nakayama, H. Miyamoto, T. Inoue, M. Senda, K. Hirata, M. Kosaki, N. Shibata, and M. Kuzuhara, *IEEE Trans. Microw. Theory Tech.* **52**, 2536- 2540 (2004).
3. S. Karmalkar, J. Deng, M.S. Shur, and R. Gaska, *IEEE Electron Dev. Lett.*, **22**, 373-375 (2001).
4. S. Karmalkar, M.S. Shur, and R. Gaska, in *Wide Energy Bandgap Electronic Devices*, edited by Fan Ren and John Zolper, World Scientific (2003), ISBN 981-238-246-1.
5. V.O. Turin and A.A. Balandin, *J. Appl. Phys.* **100(5)**, 054501 (2006).
6. Synopsys, Inc., *Sentaurus Device X-2005.10*, (2005).
7. S. Mil'shtein, *Microelectronics Journal*, **36**, 319-322 (2005).

International Journal of High Speed Electronics and Systems
Vol. 17, No. 1 (2007) 25–28

LOW TEMPERATURE ELECTROLUMINESCENCE OF GREEN AND DEEP GREEN GaInN/GaN LIGHT EMITTING DIODES

Y. Li, W. Zhao, Y. Xia, M. Zhu, J. Senawiratne, T. Detchprohm, and C. Wetzel

Future Chips Constellation, Rensselaer Polytechnic Institute, Troy, NY 12180, U.S.A.
Department of Physics, Applied Physics and Astronomy, Rensselaer Polytechnic Institute, Troy, NY 12180, U.S.A.
liy10@rpi.edu

Electroluminescence of GaInN/GaN multiple-quantum-well (QW) light-emitting diodes emitting in the green spectral region is analyzed at variable low temperature. Spectra are dominated by QW emission at RT throughout 7.7 K. Below 150 K, donor-acceptor pair recombination appears that must be assigned to residual impurities in either the barriers or the p-layers. The current-voltage behavior reveals shunt paths that carry up to several mA at low bias voltages. Below 20 K those paths are frozen out, but the device still emits predominantly from the QW. The peak energy exhibits a blue shift from RT to 158 K followed by a red shift from 158 K to 7.7 K. The deeper binding energy at low temperatures can hardly be affect by the injection current indicating that saturation of low-density states cannot be responsible for the transition between red and blue shifts.

Keywords: GaInN LED, external quantum efficiency, quantum confined Stark effect

1. Introduction

The radiative recombination processes in group-III nitride heterostructures remain the subject of continued investigation in order to increase performance of high brightness light emitting diodes (LEDs). In particular, green LEDs employing active regions of GaInN/GaN quantum wells (QWs) reveal a strong drop in emission power performance when the emission wavelength is being extended beyond 500 nm. Also the external quantum efficiency (EQE) shows a strong dependence on the drive current and temperature. It is found that for typical operation current, EQE is only a fraction of its maximum value at low current. A detailed spectroscopic analysis is therefore warranted to identify the limiting mechanism and to enhance performance in next generation solid state lighting.

2. Experimental Procedures

Three pseudomorphic GaInN/GaN multiple quantum well samples A, B, and C of nominally identical structure and InN fractions have been grown by metal-organic vapor phase epitaxy c-plane sapphire.[1,2] Five periods of GaInN/GaN QWs with 3 nm are grown between the carrier injection layers of p-type AlGaN and n-GaN. The optical properties were studied by electroluminescence (EL) at variable low temperatures from RT to 7.7 K. The temperature is that of the sample holder, and may be somewhat lower than the actual junction temperature. The EL spectra are recorded as a function of current from 0.1 mA to 30 mA by a spectrometer. The resulting data was scaled to that of quantitative power measurements at room temperature using an integrating sphere. In related studies, sample A exhibits the highest overall quantum efficiency, while sample C shows the lowest.[3] The performance of sample B lies in between and is the subject of more detailed investigation here.

3. Results and Discussion

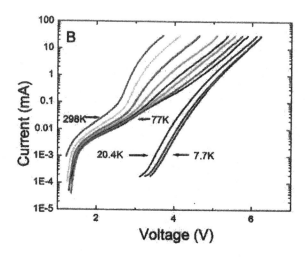

Fig. 1. Temperature dependent I-V characteristics of Sample B. The characteristics changes dramatically below 20 K, but carrier injection and light emission is still possible. There is little variation seen in Samples A and C.

The current-voltage characteristics are shown in Fig.1. The turn-on voltage, extrapolated from the exponential part of the current behavior, increases from 3.75 V at RT to 6.25 V at 7.7 K. The serial resistance is strongly temperature dependent while a constant shunt resistance is found below 0.1 mA. Carrier transport does not freeze out even at the lowest measured temperatures, yet, below 20 K, the diode switches to an entirely different transport behavior. All shunt paths are frozen out and the device behaves like an ideal diode at much lower current levels (~1/1000 that or RT). Electroluminescence is also active at such low temperatures independent of the drive current. This indicates that Joule heating due the current injection cannot be the reason for the EL. Instead, an efficient hot carrier injection mechanism needs to be considered.

The EL spectra of the identical device are shown in Fig. 2 under variable injection current from 0.1 mA to 30 mA (left panel RT, right panel 7.7 K). At temperatures below 150 K, donor-acceptor pair recombination in A with phonon replica in B appears.

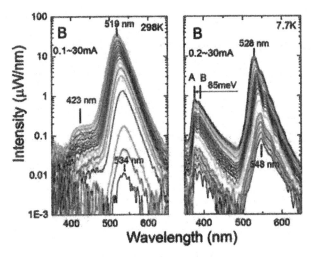

Fig. 2. Spectra of the electroluminescence at RT and at 7.7 K. The injection current varies from 0.1 mA to 30 mA. At low temperature, additional donor-acceptor pair recombination is observed.

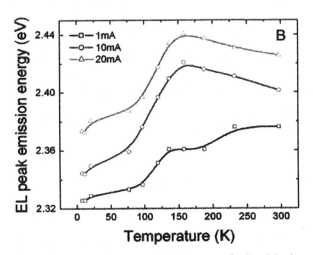

Fig. 3. The EL peak emission energy as a function of temperature for three injection current 1mA, 10 mA and 20 mA of sample B.

Its location does not vary with injection current. Its origin is tentatively assigned to either the GaN barriers or possibly the p-type layers. The origin is attributed to residual impurities in the barriers or possibly Mg acceptors in the p-layers.[4] Furthermore, multiple thickness interference fringes appear in the main QW emission at low temperature. For increasing current, the emission peak shifts to shorter wavelengths. This blue shift can be explained by a screening of the internal electric field reducing the quantum confined Stark effect (QCSE). Portions thereof might also be due to band filling at higher carrier injection. A conclusive decision cannot be drawn from this data.

It is notable that the peak emission wavelength at low temperature is longer than at RT. The peak emission energy as a function of temperature for three injection current levels is shown in Fig. 3. A very distinctive behavior can be seen for temperatures above and below 150 K. The higher temperature part can well be described by a temperature dependent bandgap energy typically described by the Varshni-formula.

Below 150 K, emission originated in deeper bound states. A frequently cited process could be carrier localization in levels of lower density-of-states.[5] To test for a density-of-state issue, we varied the injection current from 1, over 10, to 20 mA. An overall shift in emission energy towards shorter wavelength is observed, however, there is little or no effect in the behavior relative to the above 150 K part. This makes it very difficult to describe any wavelength shift as part of a variable filling of low density-of-states regions. Under such an assumption, under variable injection conditions, levels should not maintain their respective energy separations.

4. Conclusions

As part of a correlation of LED performance in terms of the quantum efficiency with defect related features, we analyzed the low temperature behavior of a green LED die performing with intermediary efficiency in the 520 – 540 nm range. At temperatures below 150 K, a donor-acceptor pair transition is identified that must be assigned to residual impurities either within the barriers or p-type layers. A large blue shift of the QW emission followed by an even stronger red shift of the main QW emission is observed, as the temperature decreases. While the bandgap change with temperature can account for the blue shift, the red shift is tentatively assigned to deeply bound levels that can not be saturated even at an injection current of 20 mA. EL operation is observed at temperatures as low as 7.7 K when measured at the sample holder that cannot be attributed to electrical heating of the sample. Instead a hot carrier or impurity band transport mechanism must account for carrier injection at such low temperatures.

References

1. C. Wetzel, T. Salagaj, T. Detchprohm, P. Li, and J.S. Nelson, GaInN/GaN Growth Optimization for High Power Green Light Emitting Diodes. Appl. Phys. Lett. **85**(6), 866-8 (2004).
2. C. Wetzel and T. Detchprohm, Development of High Power Green Light Emitting Diode Chips, MRS Internet J. Nitride Semicond. Res. 10, 2 (2005).
3. W. Zhao, Y. Li, T. Detchprohm, and C. Wetzel, The Quantum Efficiency of Green GaInN/GaN Light Emitting Diodes. Phys. Stat. Sol. (c) 4, 9 (2007).
4. K. Kim, J.G. Harrison, Critical Mg Doping on The Blue-light Emission in p-type GaN Thin Films Grown by Metal–Organic Chemical-Vapor Deposition, J. Vac. Sci. & Technol. A, 21(1), 134-139 (2003).
5. P.G. Eliseev, P. Perlin, J.Y. Lee, and M. Osinski, "Blue" Temperature-Induced Shift and Band Tail Emission in InGaN-Based Light Sources, Appl. Phys. Lett. **71**, 569-571 (1997).

International Journal of High Speed Electronics and Systems
Vol. 17, No. 1 (2007) 29–33
© World Scientific Publishing Company

SPATIAL SPECTRAL ANALYSIS IN
HIGH BRIGHTNESS GaInN/GaN LIGHT EMITTING DIODES

T. Detchprohm[1,2], Y. Xia[1,2], J. Senawiratne[1,2], Y. Li[1,2], M. Zhu[1,2], W. Zhao[1,2], Y. Xi[1,2],
E.F. Schubert[1,2,3], and C. Wetzel[1,2]

[1]*Future Chips Constellation,*
[2]*Department of Physics, Applied Physics and Astronomy*
[3]*Department of Electrical, Computer, and Systems Engineering,*
Rensselaer Polytechnic Institute, Troy, NY 12180, U.S.A.
detcht@rpi.edu

We analyze GaInN based light emitting diodes emitting in the range of 400-550nm using a new intensity-quantitative spectroscopic cathodoluminescence mapping method. Spectroscopic information of arbitrary sample locations is generated from sequences of quantitative image scans. From the temperature dependence of the intensity, we derive thermal activation energies of the dominant loss processes. Those compare well with the hole binding energies in the piezoelectric and quantized quantum well structures.

Keywords: GaInN LED, cathodoluminescence, electroluminescence, UV, blue and green

1. Introduction

GaInN based nitride semiconductors are widely implemented as common materials for highly efficient light emitting devices such as laser diodes (LD) and light emitting diodes (LED). With their high power and highly efficient optical power output, the materials are the most promising candidate for solid state lighting application. To further improve the light output performance of the device, there are several approaches under investigation worldwide such as (1) reducing threading dislocation density in the device, (2) improving internal quantum efficiency by minimizing piezo field across the quantum wells in active region and (3) increasing light extraction efficiency through special geometrical arrangement, e.g., by using photonic crystals[1], omnidirectional reflectors[2], or chip shaping[3].

However, prior to proceeding to these approaches, there are still many fundamental characteristics of the semiconductor materials that are not fully understood. Here we have investigated the emission distribution behavior of the current high brightness GaInN based LED covering peak wavelengths from 400 to 550nm via several spectroscopy techniques, i.e., electroluminescence (EL) and cathodoluminescence (CL) spectroscopy.

2. Experimental Procedures

2.1 *Sample Structures*

There are three different types of GaInN LED samples used in this study which are hereafter assigned as UV, Blue and Green LED per their emission color. Each sample is a single uncapped LED die mounted on a TO-46 header by silver epoxy. Gold wires are used to bond n and p contacts to corresponding electrodes on the headers. Each sample consists of the active regions of multiple quantum wells of GaInN quantum wells (QWs) and GaN barriers embedded between 3 μm thick n-GaN and p layers of 50 nm thick $Al_{0.05}GaN_{0.95}N$ and 0.2 μm thick p-GaN. The indium contents of 3 nm thick GaInN quantum well layer in UV, Blue and Green LEDs are approximately 0.07, 0.15 and 0.25, respectively. All samples are epitaxially grown on (0001) sapphire wafers by metal organic vapor phase epitaxy. After device fabrication, the sapphire wafers are thinned to 100 mm before being separated into an individual die. The die size is 350 μm by 350 μm with an effective mesa size of 250 μm by 250 μm. A Ni/Au transparent contact is formed as ohmic contact on mesa area which is thereafter coated by a 0.1 μm thick SiO_2 passivation layer. The optical output at 20 mA are 0.8, 2.7 and 1.2 mW and the corresponding peak wavelengths are 405, 451 and 521 nm for UV, Blue and Green LEDs, respectively. Since there are blue shifts of these peak wavelength as much as 3, 33 and 50 meV compared to those at the forward current of 1 mA for UV, Blue and Green LEDs, respectively, the emission efficiency of GaInN QWs in all sample is influenced by the piezo field.

2.2 *Cathodoluminescence Spectroscopy*

The LED headers are mounted to the cold stage capable of cooling the sample from room temperature to liquid Helium temperature. A photomultiplier (PMT) detector is employed to quantify CL intensity in photon counting mode. The electron probing current, acceleration voltage and diameter of electron beam are 600 pA, 10kV and 0.2 μm, respectively. The scanning area dimensions on the LED mesa are 37 μm by 37 μm. The CL images with a single frame resolution of 512 pixels by 512 pixels are produced via photon counting mode in order to digitally and quantitatively store the intensity information. Each pixel covers the area of 70 nm by 70 nm. The exposure time for each pixel is 32 μsec. The samples' temperatures are varied from 7.8 K to room temperature.

Here we introduce a new approach to obtain quantitative spectroscopic CL images by employing a photon counting mode. In contrast to conventional CL images, resulting data is quantitative in intensity and not distorted by brightness and contrast settings. For spectral resolution, a cinematic sequence of photon-counting-mode CL images are collected. Here we use a wavelength span of 150 nm around the LED peak emission, stepping detection wavelength in increments of 1 nm for every frame By computerized image processing, areas of interest are analyzed and combined to entire spectra. Here we consider the whole area averages and specific dark and bright spots with the images. The whole area-averaged spectrum is comparable to a conventional non-imaging CL spectrum. The size of the selected spots is 700 nm by 700 nm. The details of the spectra from such area are exhibited and discussed hereafter. We also measure and compare the typical CL spectra under the same conditions before and after acquiring the sequence of CL images in order to ensure that there is no degradation introduced to the active region during such image acquisition. While in a conventional CL method, techniques of stationary e-beam and/or high SEM magnification to confine the size of scanning area is introduced to study spatial spectral characteristics. Since these techniques require the electron beam to stay in the area of interest for much longer time to collect a CL spectrum, the degradation of a sample develop chronologically as a gradually decrease in the CL intensity regarding exposure of such sample area to the high energy electron beam.

3. Results and Discussion

Intensity and wavelength distribution of CL spectra

By analyzing the CL images, we have found that the photon energies at peak wavelength derived from the selected high and low CL intensity area spread within 30, 80 and 40 meV for UV, Blue and Green LEDs, respectively as shown in Fig. 1. Especially in UV LED, the energy spread is less than 20 meV at 188K and lower. This variation is relatively small and less than 0.5% of the average photon energy in each sample. Strong blue shift in peak wavelength and increase of peak emission intensity are both observed as ambient temperature decreases from 298 and 78K for all samples. When the sample temperature decreases further, a red shift in peak wavelength is observed at temperature below 78K and 30K in Blue and Green LEDs, respectively. The mechanism of the red shift is under investigation.

Though the uniform photon energies at peak wavelength of these samples reflect excellent homogeneity of indium composition of the GaInN QWs on a sub micron length scale, the peak intensities of the selected area vary as much as 25-50% of the average intensity value in each sample. These intensity distributions remain valid for the whole sample temperature range. For Blue and Green LEDs, there is a clear trend within each sample that CL spectra reveal with lower photon energies at peak wavelength in lower emission intensity area while CL spectra with the higher photon energies are detected for higher emission intensity area as shown in Fig. 1. Moreover, the high intensity area maintain at higher intensity than that of the low intensity area for all sample temperature in Blue and Green LEDs. But, there is no particular trend observed in the UV LED in such manner. This trend could not be explained by the QWs' thickness variation where the emission intensity would be distributed around a specific photon energy of which the well thickness is optimized. The excess indium composition around the thread dislocations acting as non recombination centers could not also be counted for this behavior since the emission peak wavelengths are quite uniform. So this trend manifests strong dependence on the quantum-confined Stark effect (QCSE) in the quantum wells and suggests localization of piezo field in the sub micron level in Blue and Green LEDs.

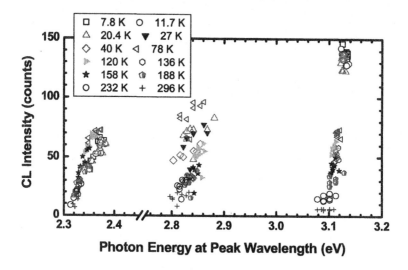

Fig. 1. Temperature dependent CL emission intensity as a function of photon energy at peak wavelength from GaInN based LEDs.

Furthermore, the activation energies of non-radiative recombination process derived from the CL peak intensity as a function of temperature are 29, 61 and 76 meV for UV, Blue and Green LEDs, respectively. There is no discrepancy of the activation energies among the high and low intensity area for all samples. In search for a mechanism that could explain this wavelength dependence of the thermal quenching behavior, we compare those values with the results of relevant bandstructure calculations of 3 nm QWs[4]. We find that these values close correspond to the effective hole binding energies in the piezoelectric QW. For LEDs emitting at the respective wavelengths, we calculate hole binding energies of 36, 63 and 92 meV. For increasing InN-fraction, the valence band offset indeed increases, while size quantization in the well and increasing piezoelectric polarization reduce the effect somewhat. These findings suggest, that the loss of thermally excited holes escaping from the GaInN QW to the valance band of GaN barrier could be an efficiency limiting factor. The CL peak intensities at room temperature are only 4, 20 and 34% of the maximum CL at low temperature for UV, Blue and Green LEDs, respectively. These values should be reasonable for these LEDs as the hole confinement improve with increasing indium content in the QWs.

4. Conclusions

We have studied sub-micron-level emission distribution of GaInN based LEDs by introducing a quantitative analysis technique of cinematically sequential CL images. From an intensity analysis of the CL temperature dependence, we find a thermal quenching behavior that can be explained by the ionization of holes from the QWs of the active region.

References

1. J. J. Wierer, M. R. Krames, J. E. Epler, N. F. Gardner, J. R. Wendt, M. M. Sigalas, S. R. Brueck, D. Li, M. Shagam, III-nitride LEDs with photonic crystal structures, *Proc. SPIE* **5739**,102-107(2005).
2. T. Gessmann, E. F. Schubert, J. W. Graff, K. Streubel, and C. Karnutsch, Omnidirectional Reflective Contacts for Light-Emitting Diodes, *J. IEEE Electron. Dev. Lett.* **24**(10), 683-685(2003).
3. V. Zabelin, D. A. Zakheim, and S. A. Gurevich, Efficiency Improvement of AlGaInN LEDs Advanced by Ray-Tracing Analysis, *IEEE J. Quan. Electron.* **40**(12), 1675-1686(2004).
4. C. Wetzel and T. Detchprohm, Development of High Power Green Light Emitting Diode Chips, *MRS Internet J. Nitride Semicond. Res.* **10**, 2 (2005).

International Journal of High Speed Electronics and Systems
Vol. 17, No. 1 (2007) 35–38
© World Scientific Publishing Company

SELF-INDUCED SURFACE TEXTURING OF AL$_2$O$_3$ BY MEANS OF INDUCTIVELY COUPLED PLASMA REACTIVE ION ETCHING IN CL$_2$ CHEMISTRY

PAOLO BATONI*

Department of Electrical and Computer Engineering & Center for Optoelectronics and Optical Communications, The University of North Carolina at Charlotte, 9201 University City Boulevard, Charlotte, NC 28223-0001, USA
pbatoni@uncc.edu

EDWARD B. STOKES

Department of Electrical and Computer Engineering & Center for Optoelectronics and Optical Communications The University of North Carolina at Charlotte, 9201 University City Boulevard, Charlotte, NC 28223-0001, USA
ebstokes@uncc.edu

TRUSHANT K. SHAH

Department of Electrical and Computer Engineering, The University of North Carolina at Charlotte, 9201 University City Boulevard, Charlotte, NC 28223-0001, USA
tkshah@uncc.edu

MICHAEL D. HODGE

Department of Electrical and Computer Engineering & Center for Optoelectronics and Optical Communication, The University of North Carolina at Charlotte, 9201 University City Boulevard, Charlotte, NC 28223-0001, USA
hodge@uncc.edu

THOMAS J. SULESKI

Department of Physics and Optical Science & Center for Optoelectronics and Optical Communications The University of North Carolina at Charlotte, 9201 University City Boulevard, Charlotte, NC 28223-0001, USA
tsuleski@uncc.edu

In this work we investigate a pseudo-random surface texturing technique of sapphire by means of inductively coupled plasma reacting ion etching in chlorine chemistry, for which no sophisticated lithographic process is required. Such a surface texturing technique, which we believe offers indicative promise for enhanced light extraction in deep ultraviolet light-emitting diodes has allowed us to texture sapphire samples having a surface larger than 1 cm^2 with controlled structures.

* 9700-B Mary Alexander Road, Charlotte, NC 28262-0848, USA.

Fabrication parameters have been characterized, and textured Al₂O₃ surfaces having submicron features, and nano-scale periodicity have been obtained. Performance, and characterization of our textured Al₂O₃ surfaces is the hinge of addition work in progress.

Keywords: LED; microstructure fabrication; diffractive optics; Al₂O₃; Cl₂ ICP etching.

Introduction

One of the most pervasive challenges involving UV LEDs (and particularly deep UV LEDs) is their low emission efficiency, which is not only restricted by the internal quantum efficiency of the active layers, and the poor conductivity of the semiconductor cladding, but also by the light extraction efficiency at the bounding surfaces. Attempts to improve the latter in longer wavelength light emitting devices (LEDs) include photonic crystals[2], surface texturing[3], and sub-wavelength antireflection gratings[4]; in addition, micro-lens array have been employed to ameliorate the light extraction efficiency in deep UV LEDs[5].

In this work we investigate a pseudo-random surface texturing technique of sapphire (Al₂O₃) by means of inductively coupled plasma (ICP) reacting ion etching (RIE) in chlorine chemistry (Cl₂), for which no sophisticated lithographic process is required. Such a surface texturing technique, which we believe offers indicative promise for enhanced light extraction in deep UV LEDs, has allowed us to texture Al₂O₃ samples having a surface larger than 1 cm² with controlled structures. Fabrication parameters have been characterized, and textured Al₂O₃ surfaces having submicron features, and nano-scale periodicity have been obtained. Performance, and characterization of our textured Al₂O₃ surfaces is the hinge of additional work in progress.

Self-induced Surface Texturing

Test structures having a surface area larger than 1-cm² were first obtained by dicing Al₂O₃ substrates; successively, they were cleaned in hydrochloric acid (HCl) for 20 minutes at 65°C. These test structures were then spin-coated with a 1.5-μm thick layer of Microposit 1813 positive photoresist (photoresist), and patterned with an UV photolithographic process controlled by a HTG Mask Aligner; next, patterns were developed with Microposit 354 positive photoresist developer in a Solitec 1100 Spray Developer system, and hard-baked in a convection oven at a temperature of 115 °C for 20 minutes. Successively, the test structures were systematically etched by means of ICP in a Cl₂, cleaned in oxygen plasma, and etch depths were finally observed and measured with a JEOL JSM 6480 scanning electron microscope.

Results and Discussions

Our etching experiments in the chlorine ICP resulted in pseudo-random textured surfaces of which we can control both periodicity, and feature size. As shown in Figure 1, textured Al₂O₃ surfaces larger than 1-cm² having submicron features, and nano-scale periodicity have been obtained; good uniformity and profile control have been demonstrated as well.

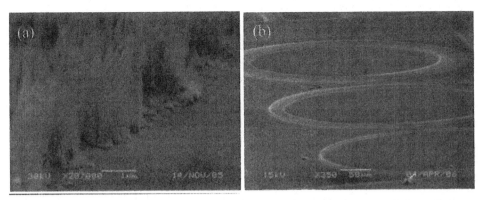

Fig. 1. (a) Detail of textured Al$_2$O$_3$ surface, and (b) detail of structure textured on textured Al$_2$O$_3$ surface.

In Figure 2 we compare the percent transmission of one of our textured sapphire structures having submicron features and periodicity in the nano-scale range, and that of a untreated sapphire sample.

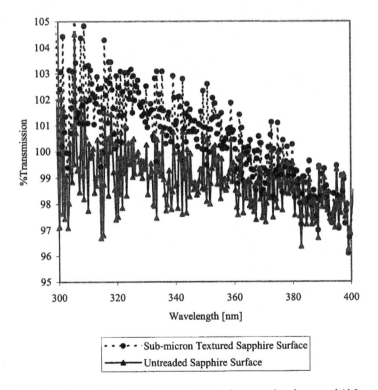

Fig. 2. Comparison of percent transmission curves obtained for textured, and untreated Al$_2$O$_3$ samples.

Experimental results suggest that our textured sapphire samples show an enhanced percent transmission of approximately 3% for wavelengths shorter than 365nm. Future work focuses on refinements of nanofabrication techniques, deposition of AlGaN active layers directly onto textured sapphire host substrates, and characterization of fabricated device performance.

Acknowledgments

This work was supported by US Army RDECOM grant # W911NF-05-2-0053. The authors would also like to acknowledge the contributions of Dr. Stephen Bobbio at UNC Charlotte and Dr. Michael Wraback from the U.S. Army Research Laboratory.

References

1. Carrano, J., and A. Khan, M. Kneissl, N. Johnson, G. Wilson, R. DeFreez, *Ultraviolet Light*, OE Magazine, June 2003, pp. 20-3.
2. H.-Y. Ryu, J.-K. Hwang *et al.*, "Enhancement of light extraction from two-dimensional photonic crystal slab structures," IEEE J. on Selected Topics in Quantum Electronics **8**, 231-237, (2002).
3. I. Schnitzer, E. Yablonovitch *et al.*, "30% external quantum efficiency from surface-textured, thin-film light emitting diodes," Appl. Phys. Letters, **63**, 2174-2176, (1993).
4. K. M. Liew, The development of 2D orthogonal polynomials for vibration of plates, PhD. thesis, National University of Singapore, Singapore (1990).
5. Y. Kanamori, M. Ishimori, and K. Hane, "High Efficient Light-Emitting Diodes With Antireflection Subwavelength Gratings," IEEE Photonics Tech. Letters, 14, 1064-1066, (2002).. Khizar, Z.Y. Fan, *et al.*, "Nitride deep-ultraviolet light-emitting diodes with microlens array," Appl. Phys. Letters **86**, 173504 (2005).

International Journal of High Speed Electronics and Systems
Vol. 17, No. 1 (2007) 39–42
© World Scientific Publishing Company

FIELD AND THERMIONIC FIELD TRANSPORT IN ALUMINIUM GALLIUM ARSENIDE HETEROJUNCTION BARRIERS

D.V. MORGAN and A. PORCH

School of Engineering, Cardiff University, Cardiff CF24 3AA, U.K.
morgandv@cf.ac.uk, porcha@cf.ac.uk

A simplified model of electron transport by tunneling within a GaAs/AlGaAs/GaAs heterojunction is developed. The model is applied specifically to tunneling through a triangular barrier formed by the compositional grading of the AlGaAs region, but can in principle be extended to a range of barrier geometries encountered at heterojunction or metal/semiconductor interfaces. The experimental data for the current-voltage characteristics obtained for a range of temperatures from 77 K to 273 K are used to test the functional dependence obtained from calculations. Good agreement has been obtained between theory and experiment, thus confirming the usefulness of the simple model for device evaluation.

Keywords: thermionic field emission, device modeling

1. Introduction

This paper develops a simple analytical model for the study of potential barriers at heterojunction interfaces in semiconductor device structures. The model considers a simple triangular barrier formed by a graded $Al_xGa_{1-x}As$ layer, of nominal thickness 50 nm, sandwiched between two n$^+$ doped GaAs regions. The composition x is varied between 0 to 0.3, yielding a barrier height E_D of around 0.25 eV. The simple analytical approach[1] is an easier alternative to the rigorous model of Padovani and Stratton[2] but is limited to investigation of the functional dependence of the terminal characteristics.

2. Theory, Results and Discussion

The present approach has been outlined in previous publications[1,3]. Fig. 1 shows the conduction band of the triangular barrier under reverse bias. Also shown in Fig. 1 is a schematic diagram of the tunneling probability though a triangular barrier $T(E_D - E)$, plotted as a function of energy $(E_D - E)$ measured from the top of the barrier, and the Boltzmann distribution $B(E)$ from the bottom of the conduction band (where, by definition, $E = 0$). Defining $g(E)$ as the density of states function, the contribution to the tunneling current density $J_{TF}(E)dE$ associated with electrons of energies between E and $E + dE$ is proportional to the product $g(E) \cdot B(E) \cdot T(E_D - E)$, so that

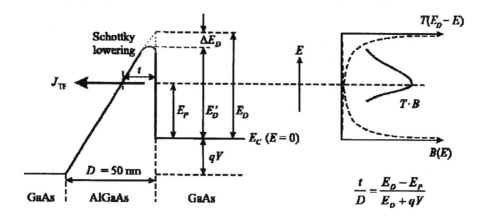

Fig. 1. Schematic diagram of the triangular barrier of the conduction band under reverse bias, introducing the effects of Schottky lowering. The quantity ΔE_D is the Schottky lowering of the energy barrier E_D. Shown on the right hand figure are the tunneling coefficient $T(E_D - E)$ (with no Schottky lowering), the Boltzmann factor $B(E) = \exp(-E/kT)$ and their product as a function of energy E from the bottom of the GaAs conduction band (where $E = 0$), up to the energy E_D of the AlGaAs interface.

$$J_{TF}(E)dE \propto g(E)\exp\left(-\frac{E}{kT}\right)\exp\left(-\frac{c}{qF}(E_D - E)^{3/2}\right)dE \,, \quad c = \frac{8\pi}{3}\left(\frac{2m^*}{h^2}\right)^{1/2} \quad (1)$$

where F is the electric field and m^* the effective electron mass in the AlGaAs barrier region. Assuming that all the transport takes place at $E = E_P$, differentiation of Eq. (1) for a constant density of states function for a simple triangular barrier yields for E_P

$$E_P = E_D - \left(\frac{2qF}{3ckT}\right)^2 \quad (2)$$

i.e.
$$J_{TF} \propto \exp\left(\frac{4q^2F^2}{27c^2(kT)^3}\right) \quad (3)$$

The analysis can be corrected for Schottky lowering by noting that its inclusion reduces E_P by an amount $\beta\sqrt{F} = 1.09\times10^{-5}\sqrt{F}$ eV, also reducing the effective barrier height by the same amount. The term $E_D - E_P$ is approximately unaffected by Schottky lowering, whereas E_P is modified according to $E_P = E_D - \alpha F^2 - \beta\sqrt{F}$; hence Eq. (3) reduces to

$$J_{TF}(E_P) \propto \exp\left(\frac{4q^2F^2}{27c^2(kT)^3} + \frac{\beta\sqrt{F}}{kT}\right) \quad (4)$$

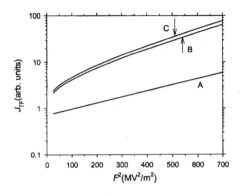
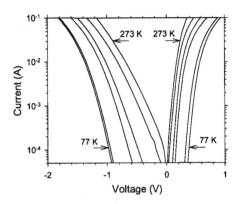

Fig. 2. The field dependence of the current density calculated at 273K assuming constant density of states with and without Schottky lowering (A and B, respectively), and variable density of states with Schottky lowering (C).

Fig. 3. Forward and reverse bias current-voltage characteristic shown on a semi-logarithmic plot for temperatures of 77, 120, 165, 195, 245 and 273K.

The effect of Schottky lowering is to enhance the device current at a fixed value of electric field F, associated with an enhanced Boltzmann factor due to the decreased value of E_P, in addition to altering the functional dependence of J_{TF} on F at low fields. Fig. 2 shows the dependence of $\ln(J_{\mathrm{TF}})$ versus F^2 calculated at 273 K; curve A assumes constant $g(E)$ and no Schottky lowering, curve B assumes constant $g(E)$ with Schottky lowering, and curve C assumes $g(E) \propto \sqrt{E}$ with Schottky lowering; in all cases the linear region is evident.

Fig. 3 shows some experimental current-voltage characteristics (I-V) for a range of temperatures from 77 K to 273 K. Fig. 4 shows the corresponding results when terminal voltage V is converted to electric field F within the AlGaAs barrier region using $F = (V + E_D / q)/D$. At the higher temperatures a linear relationship between $\log(I)$ and F^2 is observed for values of F in excess of 15 MV/m; this is the region of thermionic-field transport. At lower temperatures the peak energy E_P decreases because of its dependence on $B(E)$ and field emission takes place from the bottom of the conduction band. In these circumstances a plot of $\log(I / F^2)$ versus $1 / F$ (Fig. 5) is expected to have a linear dependence, which is indeed the case for the experimental data here.

The gradient of Fig. 5 at low temperatures can be shown to be $-cE_D^{3/2} / q$, which yields the value of $E_D \approx 0.4\,\mathrm{eV}$, which is close to the expected value of 0.25 eV. Furthermore, the gradient of the plot of Fig. 3 agrees precisely with that predicted from Eq.(3) assuming that the thickness D of the AlGaAs region is 35 nm; the anticipated value of D from the crystal growth data (assuming uniform doping) is around 50 nm, but it is quite possible that this could have been over-estimated by as much as 15 nm.

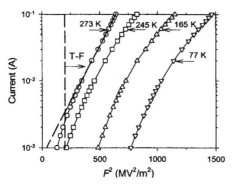
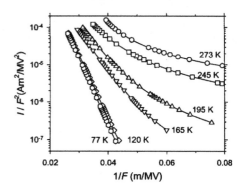

Fig. 4. A semi-logarithmic plot of current against the square of the electric field at temperatures between 77 K and 273 K. The high temperature data exhibit the linear variation predicted by Eq. (3) for thermionic field emission at large fields F (i.e. large voltages). At low fields (i.e. low voltages) the curve falls away from linearity, corresponding to thermal emission over the top of the barrier. The extent of the linear region diminishes as T is reduced as field emission then dominates, such that the data at 77K shows continuous curvature.

Fig. 5. Plots of $\log(I/F^2)$ versus $1/F$ for data taken at temperatures between 77K and 273K. The low temperature data exhibit the linear behavior predicted for field emission. In this case the Fermi-Dirac distribution (and hence the Boltzmann distribution) is attenuated and field emission from near the conduction band dominates the transport current. As expected, the high temperature data do not follow this prediction because the transport in this regime is dominated by thermionic-field emission.

3. Conclusions

In this paper the transport of electrons through the triangular conduction band of a heterojunction barrier formed from a graded AlGaAs layer has been considered. Experimental results for the current-voltage data obtained in the range 77 K to 273 K have been compared and have been shown to be consistent with thermionic-field emission (at 273 K) and field emission (at 77 K), in the latter instance as a result of the removal of more energetic electrons in the conduction band. A simplified thermionic-field model for tunneling through the barrier has been extended to include Schottky lowering and a realistic density of states function, which has been used to interpret the experimental results. The functional dependence shown in Fig. 4 based on the thermionic-field emission interpretation of the data at high temperature, and both the functional dependence and numerical value of barrier height based on the field emission interpretation of the low temperature data of Fig. 5, give confidence in the correctness of the simple model.

References

1. Morgan D.V., Board K., Wood E.C. and Eastman L.F., Physica Status Solidi (a), Vol. 72, pp. 251-260 (1982)

2. Padovani F.A. and Stratton R., Solid State Electronics, Vol. 9, pp. 695-707 (1966)

3. Morgan D.V., Porch A. and Krishna R., Physica Status Solidi (in press, 2006)

International Journal of High Speed Electronics and Systems
Vol. 17, No. 1 (2007) 43–48
© World Scientific Publishing Company

ELECTRICAL CHARACTERISTICS AND CARRIER LIFETIME MEASUREMENTS IN HIGH VOLTAGE 4H-SIC PIN DIODES

P. A. Losee, C. Li, R.J. Kumar, T. P. Chow, I. B. Bhat and R. J. Gutmann

Center for Integrated Electronics, Rensselaer Polytechnic Institute
110 8th Street, Troy, NY 12180, USA
loseep3@rpi.edu, lic6@rpi.edu, kumarj@rpi.edu, chowt@rpi.edu, bhati@rpi.edu, gutmar@rpi.edu

The key material and device parameters governing the electrical performance of high voltage 4H-SiC PiN diodes have been investigated using experimental results and numerical simulations. Reverse recovery characteristics show an increase in both carrier lifetime and anode injection efficiency at elevated temperature. Open circuit voltage decay measurements are used to estimate carrier lifetimes ($\tau \approx 0.6\mu s$ at T=25°C increasing to $\tau \approx 2\mu s$ at T=225°C) that are comparable to values measured on starting material prior to fabrication using micro-wave photoconductivity decay techniques.

Keywords: 4H-SiC PiN; Lifetime, Reverse Recovery

1. Introduction

The availability of higher quality, thicker epi-layers has led to the demonstration of SiC PiN rectifiers with blocking voltages as high as 20kV[1, 2]. Perhaps more encouragingly, large area 10kV diodes with a low forward voltage drop of 3.75V (@J_F=100A/cm^2, I_F=50A) have been presented, exhibiting sufficiently long carrier lifetimes to modulate the conductivity of lightly doped blocking layers under forward bias[3]. However, determining the carrier lifetime in SiC diodes is more difficult than in their silicon counterparts[4, 5]. In this work, drift layer carrier lifetimes in 4H-SiC PiN diodes have been extracted using electrical transient characteristics and compared with both simulated device performance and microwave photoconductivity decay (M-PCD) measurements of commercial starting material[6].

2. Device Fabrication

High voltage PiN diodes were fabricated using 110µm thick, lightly doped (N_D=6x10^{14}cm^{-3}) epi-layers grown on an n$^+$ 4H-SiC substrate (8° off-axis from 0001) purchased from CREE, Inc. A two-step anode was subsequently grown in RPI's horizontal, coldwall CVD reactor, including a 1.1µm thick p-type ($N_A \sim$ mid-10^{18}cm^{-3})

epi-layer and a 0.25μm p$^+$ cap layer (N$_A$>1x10^{19}cm^{-3}) to ensure good ohmic contact. Device isolation is achieved by mesa reactive ion etching (RIE) of the p$^+$ layers and an over etch of approximately 0.25μm is used. A 300μm wide multi-zone, implanted JTE termination is used to achieve high blocking voltage [7]. Implantations are simultaneously annealed at 1600°C in argon ambient. Al/Ni/Al and Ti/Ni/Al stacks were deposited and sintered at 1050°C for 2 minutes in argon to form the p- and n-contacts. A 1μm thick PECVD SiO$_2$ (re-oxidized at 1100°C) passivation layer is used.

Fig. 1 shows the representative best static forward and reverse current voltage (I-V) characteristics of the diodes. The experimental forward voltage drop of approximately V$_F$=4.2 – 4.5V @ J$_F$=100A/cm^2 suggests a carrier lifetime sufficient to achieve a moderate level of conductivity modulation of the lightly doped drift layer. The specific p-contact resistance measured throughout the sample (ρ$_C$ ~ 10^{-3}Ω-cm^2) contributes less than 10% of the total drop.

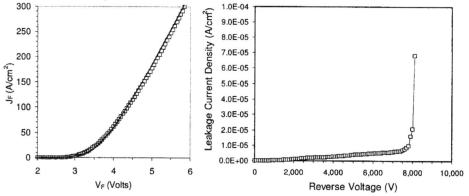

Fig. 1 Static forward (*left*) and reverse (*right*) I-V characteristics of experimental devices (measured in pulsed mode, Anode Area = 1.1x10^{-3}cm^2, reverse measured in Fluorinert oil)

3. Lifetime Measurements from Experimental Transient Characteristics

The transient characteristics of the diodes were evaluated using both inductive load reverse recovery and open circuit voltage decay measurements [4, 8]. Fig.2 shows the reverse recovery current waveforms and reverse peak current density J$_{RP}$ of a 4H-SiC PiN diode at various temperatures. The characteristics indicate an increase in carrier lifetime at elevated temperature with the extracted reverse recovery charge Q$_{rr}$ increasing from about 5μC/cm^2 at room temperature to 12μC/cm^2 at T=225°C, respectively.

Open circuit voltage decay (OCVD) measurements were used to estimate the carrier lifetime in the drift layer of the high voltage diodes [8]. The circuit used for the OCVD measurements is shown in Fig. 3, and is identical to the circuit described in Ref. 4 with the exception of the inclusion of the fast switching SiC Schottky.

Fig. 4 shows the diode voltage waveforms from OCVD measurements taken at various temperatures. The linear portion of the decay is clearly visible and can be used to estimate drift layer lifetime at the selected temperature using the relationship:

$$\tau \approx 2kt/q \, (dV/dt)^{-1} \tag{1}$$

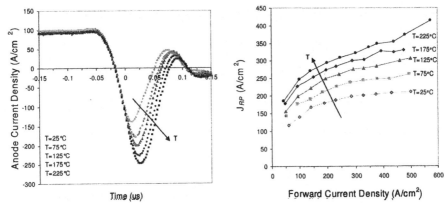

Fig. 2 Reverse recovery characteristics of 4H-SiC PiN diode at elevated temperature
(left) reverse recovery current waveform at various temperatures
(right) reverse peak current density versus forward current density
(Anode Area = $1.1 \times 10^{-3} cm^2$, Switched to V_R=60V and dJ/dt=4kA/μs-cm^2)

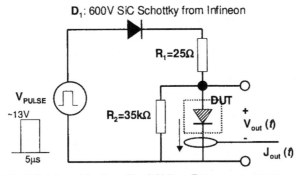

Fig. 3 Circuit used for Open Circuit Voltage Decay measurements

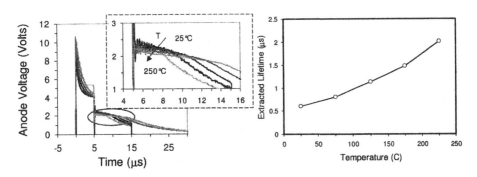

Fig. 4 Open Circuit Voltage Decay waveforms (*left*) and extracted drift layer lifetime at temperature (*right*)

Room temperature lifetimes ($\tau \approx$ 0.6μs) are found to be comparable to previously reported micro-wave photoconductivity decay (M-PCD) measurements taken on the commercial starting material prior to device fabrication [6]. As expected, and in agreement with forward I-V and reverse recovery characteristics, the extracted lifetime increased at elevated temperature (to $\tau \approx 2\mu s$ at T=225°C)

4. Comparing Simulated and Experimental Performance

MEDICI simulations were performed in order to validate the use of OCVD measurements in estimating the carrier lifetime in the fabricated 4H-SiC diodes and compare with experimental I-V characteristics. To verify the method, the drift layer lifetime was extracted from simulated OCVD characteristics under the following conditions:

(1) No locally reduced lifetime
(2) Reduced lifetime in the Anode (τ_{Anode}=1ns)
(3) Reduced lifetime at the Anode/Drift layer interface (τ_{Damage}=10ns, W_{Damage}=1μm)

The simulated OCVD waveforms for the selected conditions are shown in Fig. 5. In each of the cases shown, the extracted lifetime was within 20% of the high-level lifetime input into the simulation. In addition, the simulated waveforms yielded reasonable estimates for high-level lifetimes over the entire investigated range of τ_{HL}=0.5μs – 10μs.

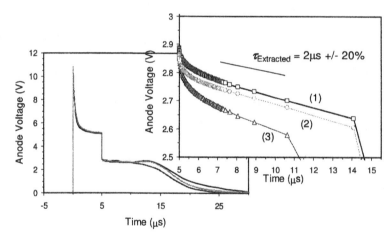

Fig. 5 Simulated 4H-SiC PiN OCVD waveforms
(1) τ_{Drift}=2μs, No interfacial damage layer
(2) τ_{Anode}=1ns, τ_{Drift}=2μs, No interfacial damage layer
(3) τ_{Drift}=2μs, 1μm/τ_{Damage}=10ns interfacial damage layer

The carrier lifetimes estimated from the OCVD measurements were incorporated in MEDICI diode I-V simulations. Fig. 6 shows good agreement between the experimental and simulated forward I-V characteristics for T=25°C, 150°C and 225°C.

The anode injection efficiency of the diodes is extracted from forward I-V characteristics using the relationship:

$$\gamma_A = I_{pA} / I_{Total} \qquad (2)$$

where I_{pA} is the hole current entering the drift layer from the anode and I_{Total} represents the total current flowing through the junction. Fig. 7 shows the extracted anode injection efficiency versus forward current density J_F. It is clear that for the anode design used in this work (along with the incomplete ionization of acceptors at room temperature), most of the current flow is made up of end region recombination at useful operating current densities (>100A/cm²). This has profound impacts on the diode's reverse recovery (both inductive and resistive load) characteristics and these methods should not be used to estimate lifetime[9]. In the presence of low injection efficiency, the excess carrier concentration near the metallurgical junction is much lower than the expected value and can lead to reduced J_{RP} and storage times t_{sd} giving erroneous values of drift layer lifetime.

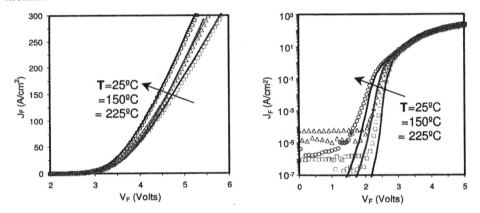

Fig. 6 Simulated (*solid lines*) and experimental (O, Δ, □) forward I-V characteristics of 4H-SiC PiN diodes Simulations include carrier lifetime extracted from OCVD measurements (measured p-contact resistance $\rho_C \approx 10^{-3}\Omega\text{-cm}^2$ included in forward I-V simulations)

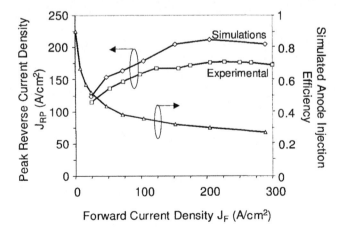

Fig. 7 Simulated anode injection efficiency and reverse peak current density versus forward current density

5. Summary

The operating characteristics of high voltage 4H-SiC PiN diodes have been investigated using experimental results and numerical simulations. The reverse recovery characteristics indicate an increase in both carrier lifetime and anode injection efficiency at elevated temperature. Open circuit voltage decay measurements are used to estimate carrier lifetimes ($\approx 0.6\mu s$ up to $2\mu s$ at elevated temperature) that are comparable to values measured using micro-wave photoconductivity decay techniques on starting material prior to fabrication. Reasonable agreement between all techniques suggests a good model, which can be used for further design optimization in the future.

Acknowledgement

This work was primarily supported by DARPA under contract number #DAAD19-02-1-0246 and used shared facilities of the National Science Foundation under Award Number EEC-9731677.

References

1. Y. Sugawara, et al, "12-19kV 4H-SiC pin diodes with low power loss", *Proceedings of International Symposium on Power Devices and ICs*, pp. 27-30, 2001.
2. M. K. Das, et al, "Latest advances in high voltage, drift free 4H-SiC pin diodes", *Semiconductor Device Research Symposium*, pp. 364-365, 2003.
3. M. K. Das, et al, " Ultra high power 10kV, 50A SiC PiN diodes", *Proceedings of International Symposium on Power Devices and ICs*, pp. 299, 2005.
4. Levinshtein et al, "Steady-state and transient characteristics of 10kV 4H-SiC diodes", *Solid-State Electronics*, vol. 48, pp. 807-811, 2004.
5. Levinshtein et al, "Paradoxes of carrier lifetime measurements in high-voltage SiC diodes", *IEEE Transactions on Electron Devices*, vol. 48, p. 1703, 2001.
6. R. J. Kumar et al, "Characterization of 4H-SiC epitaxial layers by microwave photoconductivity decay", *Materials Science Forum Vol. 483-485*, p. 405-408, 2005.
7. P. A. Losee et al, "High-voltage 4H-SiC PiN rectifiers with single-implant, multi-zone JTE termination" *Proceedings of International Symposium on Power Devices and ICs*, pp. 301-304, 2004.
8. Schlangenotto H, Gerlach W. "On the post-injection voltage decay of the p–s–n rectifiers at high injection levels" *Solid State Electronics*, vol. 15, p. 393–402, 1972.
9. B. Tien, and C. Hu, "Determination of carrier lifetime from rectifier ramp recovery waveform", *IEEE Electron Device Letters*, vol. 9, p. 553-555, 1998.

International Journal of High Speed Electronics and Systems
Vol. 17, No. 1 (2007) 49–53
© World Scientific Publishing Company

Geometry and Short Channel Effects of Enhancement-Mode n-Channel GaN MOSFETs on p and n⁻ GaN/Sapphire Substrates

W. Huang, T. Khan and T. P. Chow

Center for Integrated Electronic, Rensselaer Polytechnic Institute
Troy, NY-12180 USA
huangw2@rpi.edu

In this paper, we have fabricated and compared the performance of lateral enhancement-mode GaN MOSFETs with linear and circular geometries. Circular MOSFETs show 2 to 4 orders of magnitude lower leakage current than that of linear MOSFETs. We also studied short channel behaviors and found that they are similar to those previously reported Si MOSFET.

Keywords: GaN; MOSFET; Geometry; Short channel effects.

1. Introduction

GaN is a promising material for high power and high temperature electronics due to its wide energy band gap and large critical electric field [1]. With the development of power devices and power electronics, silicon power devices are reaching their material limits. Wide band gap materials such as SiC and GaN are therefore considered as an alternative for power electronics application especially under extreme environments. Among GaN-based electronic devices, GaN high electron mobility transistor (HEMT) attracted most of the attention making use of of the spontaneous and piezoelectric polarization charge to achieve high electron density and mobility. However many problems related to normally-on operation, gate leakage, gate lag and current collapse, poor reliability and yield still have to be solved.

GaN MOSFET is superior to GaN HEMT for its low gate leakage, large conduction band offset up to 3.6 eV [2] and positive threshold voltage for normally-off operation as a high-voltage power switching transistor. Previously we have optimized insulator/GaN interface and demonstrated normally-off MOSFETs on both p and n- type GaN epilayer on sapphire substrate with record high field-effect mobility up to 167 cm²/V·s and breakdown voltage up to 940 V [3-5].

In this paper, we describe the fabrication and characterization of enhancement-mode n-channel GaN MOSFETs with different device geometries and short channel lengths on both p and n⁻ GaN epilayer on sapphire substrates.

2. Device Structures and Fabrication

A schematic cross-section of the fabricated GaN MOSFET is shown in Fig. 1.

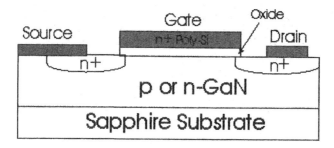

Fig. 1 Schematic cross-section of a lateral n-channel GaN MOSFETs

The lateral MOSFET was fabricated on p and n- GaN epilayer grown by metal-organic chemical vapor deposition (MOCVD) on sapphire substrates. The p epilayer thickness was 4.6 μm with Mg acceptor doping concentration of 4×10^{15} cm^{-3} whereas the n$^-$ epilayer was unintentionally doped with thickness of 3 μm. Source and drain regions were selectively implanted with a silicon dose of 3×10^{15} cm^{-2} and a maximum energy of 190 keV through 10nm SiO_2 to achieve a junction depth of 0.25 μm using PECVD oxide as mask. The wafers were then cleaned with RCA clean and diluted HCl before 600nm-thick field oxide was deposited. Annealing at 1100°C for 5 min. in N_2 was performed to activate the implanted silicon. 100nm-thick SiO_2 was deposited as gate oxide and annealed at 1000°C for 30min. in N_2 ambient [3-4]. Subsequently, 650 nm thick polysilicon was deposited and degenerately doped by $POCl_3$ to get a sheet resistance of 1 kΩ/square. After definition of the gate, 1μm thick SiO_2 was deposited to serve as interlevel dielectric. The source and drain contacts (Ti/Al and contact on source/drain and substrate) were defined using liftoff technique, and the contacts were annealed at 700°C for 30sec. in N_2 ambient using RTA. The gate contact was patterned and etched, followed by (Ti/Mo) contact metallization. The linear and circular (self-enclosed) GaN MOSFETs have channel lengths varying from 2 μm to 100 μm. Fig. 2 shows photographs of circular and linear GaN MOSFETs.

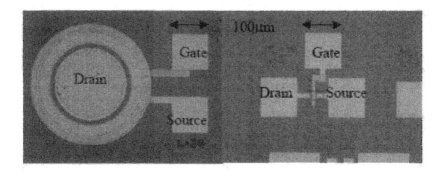

Fig. 2 Photographs of circular and linear GaN MOSFETs

3. Experimental Results

3.1 *Geometry Effects*

Figures 3 and 4 show the typical room temperature semi-log scale transfer I-V characteristics of linear and circular device with channel length 80 μm and channel width 800 μm for p and n⁻ epilayer respectively.

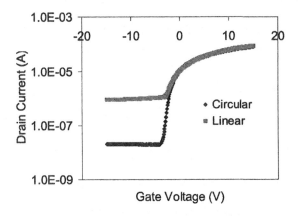

Fig. 3 Transfer I-V characteristics of linear GaN MOSFETs on p GaN with
channel length 80 μm and channel width 800 μm

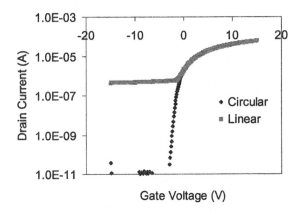

Fig. 4 Transfer I-V characteristics of linear GaN MOSFETs on n⁻ GaN with
channel length 80 μm and channel width 800 μm

The drain to source voltage was 0.1 V during the gate voltage sweep for the corresponding transfer characteristics. 2 to 4 orders of magnitude higher in off-state leakage current (1.1 μA/mm vs. 26 nA/mm and 550 nA/mm vs. < 60 pA/mm) can be seen for linear MOSFETs, on p and n⁻ GaN respectively. Circular device prevents leakage current from region other than gate region due to its self-enclosed structure, thus lower leakage current than that of linear device. GaN MOSFETs on p GaN exhibit higher leakage current than those on n⁻ GaN and we attribute this excessive leakage to epilayer defects.

Subthreshold slope for linear device is much larger than that of circular device (1.0 V/dec vs. 310 mv/dec on p GaN epilayer and 1.3 V/dec vs. 160 mV/dec for n⁻ GaN epilayer). Gate leakage current is less than 100 pA in all measurements.

3.2 *Short Channel Effects*

Figures 5 and 6 show the typical room temperature output I-V characteristics of linear devices with channel length 2 μm and channel width 20 μm for p and n⁻ epilayer respectively.

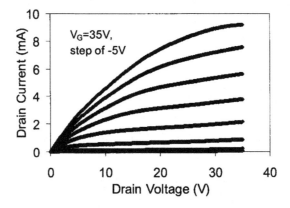

Fig.5 Drain I-V characteristics of linear GaN MOSFETs on p GaN with channel length 2 μm and channel width 20 μm

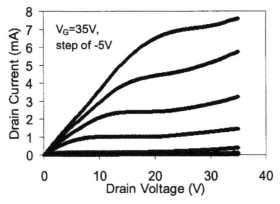

Fig.6 Drain I-V characteristics of linear GaN MOSFETs on n⁻ GaN with channel length 2 μm and channel width 20 μm

Short channel effects, such as non-saturating drain current and saturating transconductance, can be seen for both p and n- GaN epilayer MOSFETs. The maximum measured transconductance is up to 30 mS/mm for MOSFETs on n⁻ GaN and 20 mS/mm for those on p GaN. Extracted saturation velocity is 9×10^6 cm/s for MOSFETs on n- GaN and 6×10^6 cm/s for those on p GaN.

All of these short channel effects resemble strongly those observed for Si MOSFETs on bulk and SOI substrates with similar channel lengths but at 5X higher lateral channel electric field.

4. Summary

We have demonstrated enhancement-mode GaN n-channel MOSFETs with the use of annealed deposited SiO_2 for the gate oxide and ion implantation processing. Circular MOSFET with the same W/L ratio shows 2 to 4 orders of magnitude lower leakage current than that of linear MOSFETs. Short channel effects can be observed and maximum measured transconductance is up to 30 mS/mm for MOSFETs on n⁻ GaN and 20 mS/mm for those on p GaN.

Acknowledgement

The authors would like to thank the support of Center for Power Electronics Systems under NSF Award # EEC-9731677.

References

1. T. P. Chow and R. Tyagi, "Wide bandgap compound semiconductors for superior high- voltage unipolar power devices", *IEEE Trans. on Electron Devices*, Vol. 41, No. 8, pp. 1481-1483, 1994.
2. T. E. Cook, Jr., C. C. Fulton, W. J. Mecouch, K. M. Tracy, R. F. Davis, E. H. Hurt, G. Lucovsky, and R. J. Nemanich", Journal of Applied Physics, Vol. 93, pp. 3995-4004, 2003.
3. W. Huang, T. Khan, and T.P. Chow, "Comparison of MOS capacitors on N an P type GaN", Late News Paper, *Electronic Materials Conference* (2005), Accepted by *J. Electronic Materials*, 2006.
4. W. Huang, T. Khan, and T.P. Chow, "Asymmetric Interface Densities on n and p type GaN MOS capacitors", *International Conf. Silicon Carbide and Related Materials*, 2005.
5. W. Huang, T. Khan, and T.P. chow, "Enhancement-mode n-channel GaN MOSFETs on p and n⁻ GaN/Sapphire substrate", *Int. Symp. Power Semiconductor Devices and ICs*, June, 2006.

International Journal of High Speed Electronics and Systems
Vol. 17, No. 1 (2007) 55–59
© World Scientific Publishing Company

4H-SIC VERTICAL RESURF SCHOTTKY RECTIFIERS AND MOSFETS

Y. Wang, P. A. Losee, and T. P. Chow

Center for Integrated Electronics, Rensselaer Polytechnic Institute
110 8th Street, Troy, NY 12180, USA
wangy2@rpi.edu, loseep3@rpi.edu, chowt@rpi.edu

This work presents a vertical RESURF structure, which utilizes 2-dimensional depletion in semiconductor to achieve high blocking capability. This approach is then implemented into both 4H-SiC Schottky rectifiers and MOSFETs. Device characteristics are analyzed with numerical simulations and compared with both conventional Schottky rectifiers and Superjunction structure. Design-of-Experiment (DOE) is also used to optimize the trade-off between several design parameters.

Keywords: RESURF, 4H-SiC, Schottky Rectifiers, MOSFETs

1. Introduction

The wide band gap, high breakdown field and good thermal conductivity of SiC makes the material attractive for power device applications which should provide lower power dissipation and better high temperature performance than existing components. Superjunction Schottky rectifiers offer lower specific on-resistance (lower on-state losses) than other Schottky rectifiers at the desired voltage range ($\geq$1kV), but require complicated process technology, especially in SiC [1].

With the concern of lowering the on-state losses with reduced process complexity compared to the Superjunction devices, RESURF principle has been explored in both Silicon and SiC technology [2-5]. Different from the triangular field shape in 1-D structures, the field distribution in RESURF structure is much more uniform due to 2-dimensional depletion, and the device breakdown voltage is then increased effectively. In this paper, RESURF concept is introduced into vertical structures, making use of field plates in vertical direction. Fig.1 shows the schematic cross-section of a vertical RESURF Schottky rectifier. Half of the unit cell is SiC as in the conventional Schottky rectifier, and the other half is oxide. Anode metal is then not only at the top surface of the device, but also in a deep trench in the oxide. As a result, a field plate is formed with such a metal-oxide-semiconductor structure, which contributes to the 2-dimensional depletion and thus the "RESURF" effect in blocking state. Based on numerical simulations, this work details the advantage of 4H-SiC vertical RESURF rectifiers and MOSFETs.

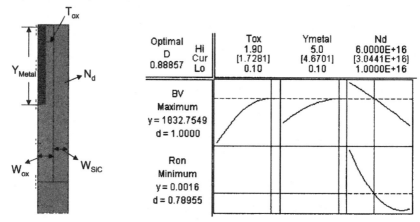

Fig. 1. Schematic cross-section of vertical RESURF Schottky Rectifier

Fig. 2. DOE optimization of 4H-SiC Vertical RESURF Schottky rectifier, W_{EPI}=10μm, T_{ox},(μm) Y_{Metal}(μm), N_D(cm^{-3}) varied y: optimized value; d: desirability

2. Simulation of 4H-SiC Schottky Rectifier

Numerical simulations have been used to verify and optimize the performance of the 4H-SiC Vertical RESURF Schottky rectifier. In the on state, current flows from the Schottky contact, through the drift region in semiconductor side, and to the substrate. Thus, only half of the device area contributes to the forward current conduction. To achieve lower specific on-resistance than the conventional Schottky rectifier, the doping concentration in the semiconductor has to be more than two times higher than that of the conventional structure.

In the reverse blocking, high voltage is applied to the cathode. Depletion region forms not only from the Schottky contact, but also from the vertical interface between oxide and semiconductor due to the field plate. Thus, 2-dimensional depletion is realized in this structure.

In order to optimize the trade-off between breakdown voltage and specific on-resistance of the device, several design parameters are considered. Fig. 2 shows a Design-of-Experiment (DOE) optimization of the oxide thickness, electrode depth and doping concentration with a 10μm thick drift layer. As expected, low specific on-resistance is achieved with high doping concentration in the drift region, and neither oxide thickness T_{ox} nor electrode depth Y_{Metal} has any effect on $R_{on,sp}$. As N_D is fixed to be 3×10^{16}cm^{-3}, high blocking capability occurs with thick oxide and deep electrode. Based on simulation results, a blocking voltage of approximately 2.1kV and a specific on-resistance of 0.85mΩ-cm^2 are obtained using W_D=10μm, N_D=3×10^{16}cm^{-3}, W_{ox}=W_{SiC}=2μm, T_{ox}=1.8μm and Y_{Metal}=5.0μm.

Fig. 3 illustrates the simulated specific on-resistance versus breakdown voltage of 4H-SiC conventional Schottky rectifiers, SJS rectifiers and selected vertical RESURF Schottky rectifier designs. It is evident that the new structures offer performance tradeoffs between that of conventional Schottky rectifiers and SJS rectifiers for voltages above 1kV.

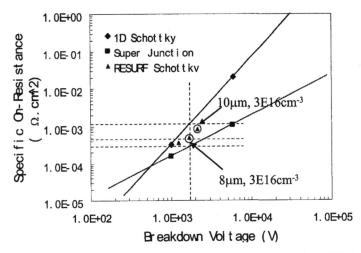

Fig. 3. Simulated performance tradeoffs of 4H-SiC Schottky, Superjunction Schottky and Vertical RESURF Schottky rectifiers (SJS: $W_P=W_N=2\mu m$)

3. Simulation of 4H-SiC MOSFET

4H-SiC Vertical RESURF MOSFETs are also evaluated with the similar structure parameters. The schematic cross-section of MOSFET using vertical RESURF technology is shown in Fig. 4. Similar with the effect of p-doped pillars in Superjunction rectifiers, the vertical field plate helps with 2-dimensional depletion of drift region when the device is in blocking state. The electric field at the interface of semiconductor and oxide (along AA') is also plotted. Because of the lateral depletion, the electric field doesn't decrease linearly as in 1-D structure. Instead, more uniform field is obtained through the drift region, which results in the improvement of blocking capability with the same drift thickness at relatively high drift doping concentration.

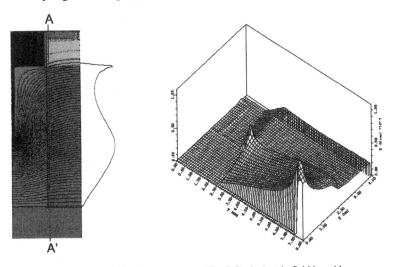

Fig. 4. Electric field distribution in reverse blocking in vertical RESURF MOSFET structure

Fig. 5. Peak electric field in oxide

Fig. 5 shows the electric field distribution in the whole vertical RESURF MOSFET structure. One concern comes from the peak electric field in the oxide. Although much better performance is obtained in semiconductor side, the electric field at sharp corner of oxide becomes to be a problem. In Fig. 5, the electric field in oxide around the bottom corner of deep electrode reaches as high as 10.5MV/cm, which is much higher than the breakdown field of deposited silicon dioxide in SiC processing. As a result, the critical field in oxide becomes another design limit.

The trade-off between peak oxide field and breakdown voltage at different oxide thicknesses and depths of gate electrode is also studied by DOE. As indicated in Fig. 6, there are two ways to reduce the peak oxide field, using either thinner oxide or shallower vertical electrode. However, both of these two changes hurt the breakdown voltage.

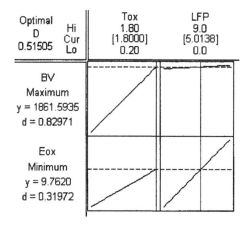
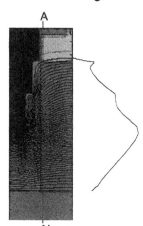

Fig. 6. DOE of blocking capability and electric field in oxide with T_{ox},(μm), and field plate length LFP(μm) varied y: optimized value; d: desirability

Fig. 7. Two-step field plate structure with BV = 1.6kV and E_{ox} = 6MV/cm

In order to achieve the minimized electric field in oxide while keeping the advantage of vertical RESURF structure, two-zone doping is considered. In such a concept, the drift region is divided into two zones, upper zone and bottom zone, and doping concentrations in these two zones are different. By adjusting the doping concentration and the zone region, it is observed that, higher doping in upper zone helps with uniformity of electric field in the drift region, but results in relatively large oxide field. On the other hand, higher doping in bottom zone is found to help reduce oxide field, but we will lose some blocking capability due to the ununiformity of electric field through the drift region.

An optimized structure is illustrated in Fig. 7. Two-step field plate is used with doping concentration of 3×10^{16}cm^{-3} in 10μm drift region. As the electric field along line AA' plotted, different from triangular electric field in 1-D structure, an opposite field distribution is achieved in semiconductor. Electric field increases from the P-N junction because of the thin oxide along first step of the field plate. Thicker oxide along second step of the field plate guarantees to make use of the upper half of drift region for blocking capability. A peak field is observed at the depth vertical electrode stops, and after that, electric field drops similar to that in 1-D structure. A breakdown voltage of 1.6kV is then obtained with such a structure, and the peak oxide field is effectively reduced to 6MV/cm which is close to that in conventional DMOS structure.

4. Implementation in SiC

In SiC, the devices start with an n⁻ epitaxial layer on highly doped n⁺ substrate. Then the vertical RESURF structure could be achieved using oxide trench etch and refill technology. This structure requires a deep trench etching through the drift region. After etching, the trench is refilled with oxide. Then, another deep trench is etched to form the field plate. Followed is metal refill, which acts as anode electrode in Schottky rectifier and gate in MOSFET. For the two-step field plate structure shown in Fig. 7, two etch steps are needed before metal refilling.

5. Summary

4H-SiC Schottky and MOSFET using vertical RESURF structure are presented. Based on the simulation results, vertical field plate helps with the improvement of trade-off between specific on-resistance and blocking capability. Peak electric field in oxide becomes another concern in these structures, which could be suppressed using two-step field plate. An optimized design shows an oxide field close to that in conventional DMOS while keeping the advantage of vertical RESURF structures.

Acknowledgement: This work was supported by the ARL/Honeywell Collaborative Technology Alliance in Power and Energy and NSF Center for Power Electronic Systems (# EEC-9731677).

References

1. L. Zhu, P. Losee and T.Paul Chow, "Design of High Voltage 4H-SiC Superjunction Schottky Rectifiers", *International Journal of High Speed Electonics and Systems*, vol.14, No.3, pp.865-871, 2004
2. J.A.Appels, and H.M.J. Vaes, "High Voltage Thin Layer Devices (RESURF Devices)", *International Electron Devices Meeting*, vol.25, pp. 238-241, 1979.
3. S.Banerjee, T.P.Chow, and R.J.Gutmann, "Robust, 1000V, 130 mΩ-cm², Lateral, Two-Zone RESURF MOSFETs in 6H-SiC", *International Symposium on Power Semiconductor Devices & ICs*, pp. 69-72, 2002.
4. C. Rochefort, and R. van Dalen, "Vertical RESURF Diodes Manufactured by Deep-Trench Etch and Vapor-Phase Doping", *IEEE Electron Device Letters*, vol. 25, pp. 73-75, Feb. 2004.
5. R. Van Dalen, C. Rochefort, and G.A.M. Hurkx, "Breaking the Silicon Limit using Semi-insulating Resurf Layers", *Proceeding of International Symposium on Power Semiconductor Devices & ICs*, pp. 391-394, 2001.

International Journal of High Speed Electronics and Systems
Vol. 17, No. 1 (2007) 61–80
© World Scientific Publishing Company

World Scientific
www.worldscientific.com

PRESENT STATUS AND FUTURE DIRECTIONS OF SiGe HBT TECHNOLOGY

Marwan H. Khater[a], Thomas N. Adam[b], Rajendran Krishnasamy[c], Mattias E. Dahlstrom[c], Jae-Sung Rieh[d],
Kathryn T. Schonenberg[a], Bradly A. Orner[c], Francois Pagette[a], Kenneth Stein[b], and David C. Ahlgren[b]

[a]*IBM T. J. Watson Research Center, 1101 Kitchawan Road, P. O. Box 218,*
Yorktown Height, New York 10598, USA
mkhater@us.ibm.com

[b]*IBM Microelectronics, 2070 Route 52, Hopewell Junction, New York 12533, USA*

[c]*IBM Microelectronics, 1000 River Street, Essex Junction, Vermont 05452, USA*

[d]*School of Electrical Engineering, Korea University, Seoul, Korea 136-701*

The implementation of challenging novel materials and process techniques has led to remarkable device improvements in state-of-the-art high-performance SiGe HBTs, rivaling their III-V compound semiconductor counterparts. Vertical scaling, lateral scaling, and device structure innovations required to improve SiGe HBTs performance have benefited from advanced materials and process techniques developed for next generation CMOS technology. In this work, we present a review of recent process and materials development enabling operational speeds of SiGe HBTs approaching 400 GHz. In addition, we present device simulation results that show the extendibility of SiGe HBT technology performance towards half-terahertz and beyond with further scaling and device structure improvements.

Keywords: SiGe HBTs; vertical scaling; lateral scaling; self-aligned structure; raised extrinsic base.

1. Introduction and Overview

The improvement in SiGe HBT transistor performance, especially the operation speed, is an essential requirement for increased bandwidth and data transfer rates in modern network communication systems. SiGe HBTs have been favored for RF/analog/mixed-signal applications owing to their advantages in transconductance, $1/f$ noise, device matching, and power performance, as compared to CMOS transistors. In addition, a monolithic integration compatibility with standard CMOS technologies and high reliability has made them an attractive and low-cost alternative to III-V HBT technologies. Recent developments in SiGe HBT transistor technology allowed operation speeds approaching 400 GHz[1-4], minimum gate delays below 3.3 ps[3-5], and enabled circuits operation at 60 GHz[6]. The cut-off frequency (f_T) and maximum oscillation frequency (f_{MAX}) of SiGe HBTs have been improved significantly by vertical and lateral scaling enabled by modern CMOS process techniques readily available for SiGe HBT and BiCMOS technologies. However, scaling of SiGe HBTs has limitations similar to those encountered in CMOS technology that can only be overcome by the advancement of new materials, process techniques, and structural innovations[3,4,7,8,9].

The performance of SiGe HBTs in recent years, compared to current InP-based HBT technologies[10-17], is shown in Fig. 1. Note that the common trend in HBT technology development is to achieve a device with balanced f_T and f_{MAX}. Aggressive vertical scaling and impurity-profile engineering of the SiGe base, enabled by modern epitaxial growth techniques, allowed SiGe HBTs to operate at f_T near 400 GHz[2,4], approaching frequencies once thought achievable only with III-V material based HBTs. In addition, as shown in Fig. 1, the capability of SiGe HBTs operation at f_T above 500 GHz was recently demonstrated by operating the device at a temperature of 4.5 K[18]. The improvement in f_T, however, trades off and limits f_{MAX} of the device due to increased parasitic capacitance and resistance caused by vertical scaling according to the approximate relation

$$f_{MAX} \approx \sqrt{\frac{f_T}{8\pi R_B C_{CB}}} \tag{1}$$

where, R_B is the total base resistance and C_{CB} is the collector-to-base capacitance. The parasitic (i.e. extrinsic) components of R_B and C_{CB} can be optimized to further improve f_{MAX} by lateral scaling and device structure modification enabled by advanced CMOS-compatible lithography and process techniques. A significant structural improvement of SiGe HBTs is the implementation of a raised extrinsic base self-aligned to the emitter, which allows reduction of R_B and C_{CB} independently[7,8]. Lateral scaling and device structure improvements enabled SiGe HBTs to operate at f_{MAX} up to 350 GHz[3].

In this paper, state-of-the-art SiGe HBTs performance is reviewed with emphasis on scaling, materials, process techniques, and structural modifications enabling the device performance improvement. In addition, future directions of SiGe HBT technology are discussed based on process and device simulations of further scaling, new process technologies, and materials implementation to further improve the device performance.

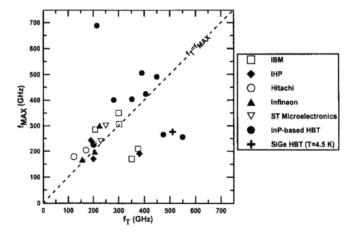

Fig. 1. Evolution of f_T and f_{MAX} of SiGe HBTs and InP-based HBTs.

2. SiGe HBT Device Structure

The schematic and SEM cross-section views of a modern SiGe HBT device are shown in Fig. 2. Combined deep and shallow trenches (DT and STI) provide device isolation. A buried subcollector and an n⁻ epitaxial layer form the collector region along with the selectively-implanted collector (SIC) pedestal. A boron-doped SiGe:C base layer is grown by non-selective UHV/CVD, and a boron-doped polysilicon raised extrinsic base is formed self-aligned to the in-situ phosphorus-doped emitter. A cobalt silicide (CoSi$_2$) formed on the raised extrinsic base polysilicon and collector reach-through serves as an ohmic low-resistance contact layer to the base and collector.

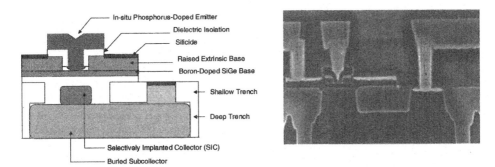

Fig. 2. Schematic and SEM cross section views of SiGe HBT with raised extrinsic base.

A significant structural improvement of SiGe HBTs is the implementation of a polysilicon raised extrinsic base self-aligned to the emitter, where the emitter-to-base spacing is determined by a spacer width, as shown in Figs. 2, rather than implanted extrinsic base[19,20]. The extrinsic base is usually doped by boron implantation. The ion implant conditions for an implanted extrinsic base have to be optimized to balance the trade-off between R_B and C_{CB}. Low R_B requires high dose and energy implant, whereas low C_{CB} requires low dose and energy implant. In other words, R_B and C_{CB} are coupled in an implanted extrinsic base. In addition, implanting the extrinsic base limits lateral scaling of the device, where the implant creates silicon interstitials which cause the base dopants to diffuse at high temperature processing[20,21]. This limits the proximity of the extrinsic base to the intrinsic base portion of the device and thus limits base and emitter scaling. On the other hand, a raised extrinsic base, which allows independent optimization of R_B and C_{CB}, minimizes both of these effects by locating the low resistance region above the intrinsic base. In this case, both the base implant defects and the intersection of the collector and base dopants in the intrinsic device are eliminated. In addition, rather than implanted, the raised extrinsic base polysilicon could be in-situ doped with boron using CVD techniques to eliminate implant defects and further improve the device performance. High boron doping levels on the order of 10^{19}-10^{21} cm^{-3} can be achieved by implantation or in-situ doping techniques to obtain polysilicon extrinsic base with sufficiently low sheet resistance.

The critical lateral scaling dimensions and components of R_B and C_{CB} for the self-aligned device structure with raised extrinsic base are shown in Fig. 3. The emitter width (W_E) is defined by lithography and a spacer width (W_S) formed in a similar fashion implemented in CMOS technology. In addition, the collector width (W_C), SIC implant width (W_{SIC}), and base silicide-to-emitter spacing (D) are also defined by lithography.

Total C_{CB} components include the intrinsic capacitance (C_{int}) and the extrinsic capacitance (C_{ext}). C_{int} is determined by the SIC doping levels and width as well as the emitter width, W_E. C_{ext} includes many components that are governed by the device structure and process techniques. The capacitance under and adjacent to the spacer (C_{SIC}) is influenced by the SIC implant width, W_{SIC}, and dopant lateral diffusion due to processing at high temperatures. The link capacitance component (C_{link}) is controlled by the link area overlap between the collector and base. The capacitance over STI (C_{STI}) is determined by the overlap area between the collector and base over STI and the STI material dielectric properties.

Total R_B components include the intrinsic base resistance (R_{int}) and the extrinsic base resistance (R_{ext}). R_{int} is determined by the intrinsic SiGe base thickness and doping levels as well as the emitter width, W_E. R_{ext} includes many components that are determined by the device structure and process techniques. The base resistance under the spacer (R_{spr}) is determined by the spacer width, W_S. The link resistance component (R_{link}) is controlled by the contact area and quality of the interface between the intrinsic and extrinsic base. The extrinsic base polysilicon resistance component (R_{poly}) is governed by the boron doping level, polysilicon thickness, and spacing D between the silicide and emitter. The silicide contact resistance component (R_{sc}) is controlled by the silicide to silicon interface properties, while the silicide resistance component (R_{sil}) is determined by the silicide type.

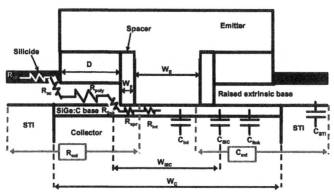

Fig. 3. Critical lateral scaling dimensions and components of R_B and C_{CB} for a self-aligned SiGe HBT device structure with raised extrinsic base.

3. SiGe HBT Technology: Current Status

3.1. *Vertical Scaling and Impurity Profile Engineering*

Vertical scaling and dopant profile optimization of the collector and SiGe base are key factors that help improve the device f_T. More specifically, the thicknesses and doping levels determine the delay times in the "*intrinsic*" device. f_T is improved as carrier transit-times and sheet resistances are reduced by thinning the collector and base layers and by boosting their doping levels.

Collector vertical scaling and dopant profile optimization are performed by controlled changes during the formation of the in-situ doped subcollector and selectively-implanted collector (SIC) region, as schematically depicted in Fig. 4. An in-situ doped subcollector layer is epitaxially grown by chemical vapor deposition (CVD). Subsequently, an n⁻ layer is epitaxially grown to separate the active device from the subcollector region. The doping level in the n⁻ layer is lower at the collector-base junction to reduce C_{CB} in the extrinsic device. However, the collector doping in the intrinsic device needs to be sufficiently high to reduce the collector resistance (R_C) in order to improve f_T. This is achieved by selective implantation of the intrinsic device using lithographic techniques, which can be properly scaled (i.e. W_{SIC}) to reduce R_C without significant impact on C_{CB} in the extrinsic device. A carefully designed SIC implant is adequate to provide a sufficient link conductance between the base and a subcollector with thickness below 0.5 μm. SIC implant energies and doses are usually in the range between 30 to 120 keV and 10^{14} to 10^{16} cm⁻², respectively.

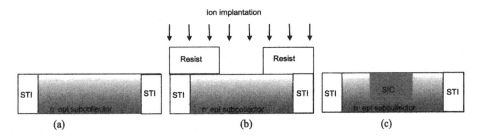

Fig 4. Schematic depiction of SiGe HBT collector formation: (a) epitaxial growth of n⁻ and subcollector layers and STI formation, (b) and (c) SIC formation using lithography.

The effect of normalized SIC total dose on C_{CB}, f_T, and f_{MAX} is shown in Fig. 5. The device SiGe base was optimized with 100% SIC total dose to obtain balanced f_T and f_{MAX} of 300 GHz[22]. As can be seen from Fig. 5, as the SIC total dose is increased to 100% to reduce R_C, C_{CB} increases by 30% (from 4.5 to 6.4 fF) leading to about 43% improvement in f_T (from 178 to 310 GHz) without a significant effect on f_{MAX}. In this case, the increase in C_{CB} is mainly due to an increase in both the intrinsic component C_{int} and the extrinsic component C_{SIC} since the implant is confined within W_{SIC}, which leads to a slight degradation in f_{MAX} by 15 GHz. In addition, no appreciable change in R_B was observed over the investigated SIC dose range. Recently, a similar technique has been implemented in the fabrication of InGaAs/InP double HBTs, where the subcollector was formed by a blanket Fe implant followed by a patterned Si implant to form the collector pedestal in order to reduce the extrinsic C_{CB}[17].

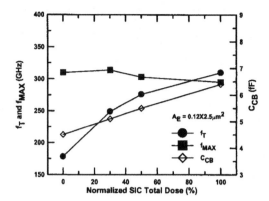

Fig. 5. C_{CB}, f_T, and f_{MAX} as a function of normalized SIC total dose.

The crystalline SiGe base layer is one of the most demanding parts of the HBT and technological challenges arise during "classical" device scaling. Vertical scaling of the SiGe base thickness and doping profile is achieved with modern low temperature UHV-CVD techniques, which enabled epitaxial growth of aggressively in-situ doped thin SiGe layers (below 50 nm) with excellent doping profile control[23-26]. Ultra-low deposition temperatures are required for a) thermal budget considerations during BiCMOS integration, b) dopant and alloy abruptness, and c) film and interface cleanliness. Typically, temperatures well below 600°C are necessary for a technology requiring a near-atomic dopant and alloy control. However, as thicknesses are reduced and dopant profiles become narrower, emitter and collector diffusion as well as base widening during subsequent activation and re-crystallization anneals severely limited further device scaling. The incorporation of carbon in the in-situ boron-doped SiGe base (i.e. SiGe:C) was found to suppress thermal boron diffusion caused by high thermal-budget processing and thus enabled continued and extendible base scaling[27]. A graded carbon doping in the base was implemented in fabricating InGaAs/InP double HBTs to reduce the base sheet and contact resistivities in order to improve f_{MAX}[10].

A typical SiGe base profile and scaling to improve f_T is shown in Fig. 6, where carbon profile is not shown. The total SiGe base thickness includes the neutral base layer and both emitter and collector intrinsic (here undoped) silicon layers (i.e. i-layer). The neutral SiGe base layer, containing the boron and carbon dopants, has a graded-Ge profile which creates a quasi electric field accelerating electrons across the base and hence improving f_T. The collector and emitter intrinsic layers allow dopant diffusion during the emitter thermal anneal to form the collector-base and emitter-base junctions. In addition, a sufficient collector i-layer thickness is critical for high quality epitaxial SiGe base layer growth otherwise degraded by residual contamination at the metallurgical collector-base interface. The neutral SiGe base width W_B determines the base transit time and requires optimization to improve f_T. For example, if W_B is reduced without adjusting the boron doping concentration, it may significantly increase R_B and degrade f_{MAX}. Therefore, efforts for minimizing W_B need to be carried out simultaneously with doping concentration increases, which is a challenging task. Two growth parameters are widely used to control W_B: diborane flow rate (i.e. boron concentration) and as-grown base width. The base profile in Fig. 6(b) was scaled down as compared to Fig. 6(a) to improve f_T by a) reducing the thicknesses of emitter i-layer, neutral base width, and collector i-layer, b) adjusting boron dose and width, and c) increasing the Ge gradient for a higher built-in quasi-static drift field in the neutral base.

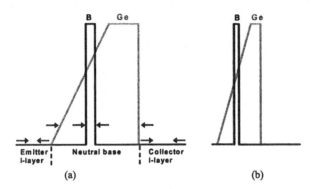

Fig. 6. Typical SiGe:C base doping profiles and scaling to improve f_T. Carbon profile is not shown.

The effect of normalized collector intrinsic layer thickness on C_{CB}, f_T, and f_{MAX} is shown in Fig. 7. The collector i-layer thickness modulates the distance between the SIC and the base layer. Thickening the collector i-layer would effectively decrease both the intrinsic and the extrinsic components of C_{CB}, unlike the case with reduced SIC total dose (Fig. 5), which affects only the intrinsic component C_{int} and the extrinsic component C_{SIC}. As a result, the increase in collector i-layer thickness would have inverse effect on f_T and f_{MAX}, which can be seen from Fig. 7. As the collector i-layer thickness is increased by 60%, C_{CB} decreases by about 20% (from 7.2 to 6 fF), with no appreciable change in R_B, leading to about 43 GHz (~27%) increase in f_{MAX} (from 159 to 202 GHz). On the other hand, such an increase in the collector i-layer thickness results in about 10 GHz decrease in f_T (from 341 to 331 GHz) caused by an increase in R_C (i.e. transit time).

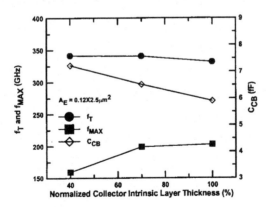

Fig. 7. C_{CB}, f_T, and f_{MAX} and as a function of normalized collector intrinsic layer thickness.

The effect of normalized as-grown base width W_B on R_B, f_T, and f_{MAX} is shown in Fig. 8. As W_B is increased by about 45%, R_B decreases by about 16% (from 42.2 to 35.4 Ω) leading to 20 GHz increase in f_{MAX} (from 306 to 326 GHz). In this case, the decrease in R_B is mainly due to a decrease in R_{int}, where the base current flows laterally through a wider W_B towards the base contact. On the other hand, f_T decreases by 24 GHz (from 303 to 279 GHz) due to an increase in the base transit time caused by reduced Ge ramp slope for wider W_B. The peak Ge percentage in the base determines the Ge ramp slope (Fig. 6) and thus affects the quasi electric field and carrier transport across the base region. As the peak Ge percentage is increased for a nominal W_B, the electric field increases leading to a reduction in the base transit time and improvement in f_T. For example, increasing the peak Ge fraction by 28% resulted in a moderate increase in f_T by ~11 GHz whereas f_{MAX} degraded by ~13 GHz due to 20% increase in R_B. f_T and f_{MAX} can be simultaneously improved with W_B reduction accompanied with an increase in the boron doping concentration, which can be achieved by increasing the diborane flow during the SiGe base growth. A two-step Ge profile in the base and boron in-situ doping of the emitter i-layer was shown to increase f_T at low currents allowing ultra-low power operation of SiGe HBT at high speeds[28].

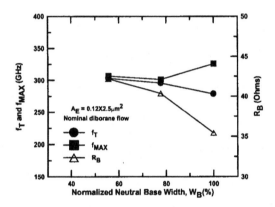

Fig. 8. R_B, f_T, and f_{MAX} as a function of normalized neutral base width (W_B).

The emitter resistance (R_E) is an additional important parameter that impacts f_T of SiGe HBTs. Recent studies have shown that a significant reduction in R_E as well as better controllability is achieved with in-situ doped emitter polysilicon compared to conventional implanted emitter polysilicon[29]. Phosphorous or arsenic in-situ doped polysilicon with high doping levels are implemented as emitter layers in modern SiGe HBT devices to improve f_T and control the device DC gain (β). Typical emitter polysilicon layer doping levels are on the order of 10^{20} cm^{-3}. The emitter polysilicon is usually deposited by rapid-thermal reduced-pressure CVD (RTCVD) for lowest thermal budget. Crystalline re-alignment, and dopant activation and out-diffusion from the emitter polysilicon are usually achieved with rapid thermal annealing (emitter RTA), in the range of 900-1000°C, which reduces R_E and forms the emitter-base junction. Cross-sectional TEM of a re-aligned emitter after activation anneal is shown in Fig. 9. Typical emitter layer sheet resistances are between 1-4 mΩ-cm. A metal emitter, made by complete silicidation of a mono-crystalline emitter (i.e. fully silicided emitter), was shown to reduce R_E and improve the device f_T and breakdown voltage[9].

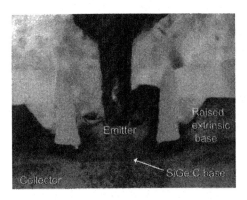

Fig. 9. Cross-sectional TEM of re-aligned emitter after activation anneal.

3.2. *Lateral Scaling and Device Structure Modifications*

SiGe HBT device performance can be further improved by lateral scaling and device structure modifications. Lithography techniques can be used, in a similar fashion implemented in CMOS technology, to reduce W_{SIC} and W_C thus minimizing C_{CB}. Furthermore, lithography and process techniques can be used to optimize the silicide-to-emitter spacing, D, and spacer width, W_S, to reduce R_B. In addition, the emitter width, W_E, in modern SiGe HBT technologies can be scaled down to sub-100 nm dimensions using advanced lithography techniques to reduce R_B and C_{CB} in order to improve the device speed.

As the SIC dimension W_{SIC} is reduced, C_{CB} decreases and R_C increases, while R_B remains approximately constant. However, the time delay product $R_C \cdot C_{CB}$, which determines the device speed, was shown to decrease with SIC lateral dimension[30]. In this case, the reduction in W_{SIC} reduces the capacitance components C_{SIC}. As a result, f_T increases due to the decrease in delay time and f_{MAX} increases due to the reduction in C_{CB}. Furthermore, the time delay $R_C \cdot C_{CB}$ could be optimized by lateral scaling of the collector width, W_C. The effect of W_C lateral scaling on f_T and f_{MAX} is shown in Fig. 10. As W_C is reduced from 0.72 μm to 0.40 μm, C_{CB} reduces by about 25% (from 12.2 to 9.15 fF) leading to an increase in f_T by 16 GHz (~5%) due to the decrease in delay time and an increase in f_{MAX} by 10 GHz (~6%). In this case, reduction in W_C reduces the capacitance component C_{link} and increases the component C_{STI}. However, C_{link} decrease is more effective in reducing total C_{CB}, since the contribution of C_{STI}, where the base and collector are separated by a thick oxide in the STI region, is negligible. A novel collector structure was shown to reduce C_{CB} by selectively undercutting the collector region to reduce the capacitance components C_{SIC} and C_{link}[4].

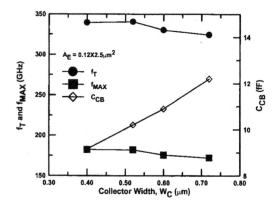

Fig. 10. f_T and f_{MAX} as a function of collector width (W_C).

State-of-the-art lithography tools employed in CMOS technology scaling to sub-100 nm dimensions are used to reduce the emitter dimension in SiGe HBTs to lower the device R_B and C_{CB}. The decrease in emitter (i.e. intrinsic device) area in effect reduces the intrinsic components R_{int} and C_{int}. In addition, reducing W_E leads to an increase in the link contact area between the base and collector leading to a decrease in R_{link} and a moderate increase in C_{link}. However, the decrease in C_{int} is more effective in reducing total C_{CB}, since the increase in C_{link} for the investigated W_E range is negligible. The effect of W_E scaling on R_B, C_{CB}, f_T, and f_{MAX} is shown in Fig. 11. Reducing W_E from 0.2 μm to 0.12 μm results in a reduction of about 9% and 12% in R_B and C_{CB}, respectively, which leads to a significant increase of 30 GHz in f_{MAX} and a moderate increase of 8 GHz in f_T. The emitter width could readily be reduced below 100 nm to further improve SiGe HBT performance. However, aggressive lateral reduction of W_E may increase R_E and R_C in the intrinsic device, which leads to f_T degradation[31]. This can be compensated for by increasing the doping levels in the emitter and collector.

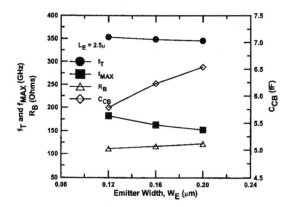

Fig. 11. R_B, C_{CB}, f_T, and f_{MAX}, as a function of emitter width (W_E).

Another technique to reduce the device parasitics is the optimization of the device layout. Two device layout configurations are shown in Fig. 12, where CBE and CBEBC represent the relative order of electrode contacts. SiGe HBT devices have traditionally adopted the compact CBE configuration as the contact resistance does not limit performance because of the availability of silicide. However, as the device speed enters the hundreds-of-GHz operation regime, the effect of parasitic resistance and capacitance becomes more significant and the device layout needs to be considered for any performance enhancement. The CBEBC configuration improves f_T and f_{MAX}, compared to the CBE configuration, due to R_C reduction and the symmetric spread of injected electrons in the collector region as well as reduction of R_B component along the silicided region. The CBEBC configuration alone improved f_T and f_{MAX} by 25 GHz and 40 GHz, respectively[2].

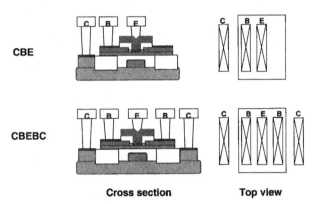

Fig. 12. Schematics of SiGe HBT device layout configurations: CBE and CBEBC.

3.3. *State-of-the-Art SiGe HBT Peroformance Path (IBM)*

Figure 13 shows the performance improvement through vertical scaling and device structure improvements for a SiGe HBT device with an emitter area A_E=0.12×2.5 μm^2. The first implementation of a self-aligned device structure with a raised extrinsic base and optimized vertical scaling[32] achieved f_T of 200 GHz and f_{MAX} of 285 GHz[7]. To improve f_T, the device vertical profile was scaled down by 1) collector vertical scaling, 2) SiGe base vertical scaling, which included thickness reduction, boron and Ge width reduction, and Ge gradient increase, 3) emitter activation anneal reduction. This resulted in an improvement in f_T to 280 GHz whereas f_{MAX} degraded to 170 GHz due to an increase in C_{CB} and R_B caused by collector vertical scaling and base layer thinning. To further improve f_T, additional collector vertical scaling to reduce R_C and an increase in peak Ge percentage (i.e. Ge gradient slope) were performed. Since these changes mainly affect the intrinsic device, f_T improved to 350 GHz while f_{MAX} remained approximately the same at 170 GHz[1]. Using the same vertical profile, the device layout was subsequently modified from CBE to CBEBC configuration, which improved f_T and f_{MAX} to 375 GHz and 210 GHz, respectively[2]. In an effort to achieve balanced f_T and f_{MAX}, the peak Ge percentage was decreased to reduce the quasi electric field in the base region in order to reduce f_T, while the base vertical profile was modified to reduce C_{CB} and improve f_{MAX}. In this case, a device with balanced f_T and f_{MAX} both of which exhibiting 300 GHz, was achieved[22]. Finally, the silicide-to-emitter spacing D was reduced to lower R_B (i.e. R_{poly}) in order to improve f_{MAX}. As a result, a device with f_{MAX} of 350 GHz, without affecting f_T of 300 GHz, was achieved[3]. f_T and f_{MAX} for this device are plotted in Fig. 14 as a function of collector current (I_C) and selected device parameters are summarized in

Table 1. As depicted in Fig. 13, vertical scaling results in a trade-off between the device performance and breakdown voltage as expected. Aggressive vertical scaling aimed to improve f_T and peak DC current gain (β) results in reduced collector-emitter breakdown voltage (BV_{CEO}) and collector-base breakdown voltage (BV_{CBO}).

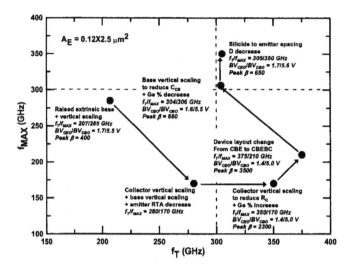

Fig. 13. State-of-the-art SiGe HBT technology performance path (IBM).

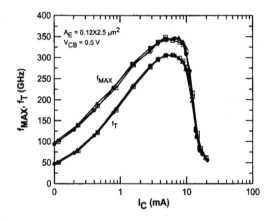

Fig. 14. f_{MAX} and f_T extrapolated from U and h_{21} at 40 GHz with -20 dB/dec slope.

Table 1. Selected device parameters of SiGe HBT with emitter area A_E=0.12×2.5 μm².

Parameter	Value
Peak f_{max}	350 GHz
Peak f_T	300 GHz
J_C @ peak f_{max} and f_T	19 mA/μm²
Peak β	650
BV_{CBO}	5.6 V
BV_{CEO}	1.7 V
BV_{EBO}	2.5 V

4. SiGe HBT Technology: Future Directions

SiGe HBT device performance can be further improved by vertical scaling, lateral scaling, and structure modifications enabled by modern process techniques and new materials developed for state-of-the-art CMOS technologies. In addition, technology simulation tools can be used to predict the impact of extended scaling and the implementation of novel materials on the device performance. In this section, we present simulation results to determine possible future directions to improve SiGe HBTs performance.

4.1. *SIC Implant Species*

Phosphorus (P) has been conventionally used as SIC implant species in modern SiGe HBTs due to its low activation thermal budget, which helps maintain high f_T. However, the high diffusion coefficient of phosphorus results in a significant lateral diffusion to the extrinsic device which leads to an increase in the extrinsic capacitance component C_{SIC}, hence degrading f_{MAX}. Other species with lower diffusion coefficient, such as arsenic (As) and antimony (Sb), can be implemented to reduce the lateral diffusion of SIC implant to maintain a lower C_{SIC}. Simulation results in Fig. 15 show the effect of phosphorus, arsenic, and antimony SIC implants on C_{CB}, f_T, and f_{MAX}. The total dose and anneal temperature were the same for all implant species. As can be seen from Fig. 15, C_{CB} is lower for As and Sb (~ 17%) compared to P leading to an increase of about 54 GHz in f_{MAX} (from 316 GHz to 370 GHz). However, a slight decrease in f_T is observed due to a decrease in the vertical diffusion of As and Sb for the same anneal temperature compared to P. We note that As and Sb SIC implants had insignificant effect on R_B.

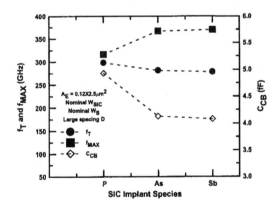

Fig. 15. Simulated C_{CB}, f_T, and f_{MAX}, for different SIC implant species: Phosphorus (P), Arsenic (As), and Antimony (Sb), for a device with large D.

4.2. *Lateral Scaling and Device Structure Modifications*

A further reduction in the extrinsic capacitance component C_{SIC} can be achieved by reducing the SIC implant width W_{SIC} using modern lithography techniques readily available for SiGe HBT technology. The simulation results shown in Fig. 16 demonstrate the effect of normalized W_{SIC} (nominal W_{SIC}=100%) scaling on C_{CB}, f_T, and f_{MAX} for a device with an emitter area A_E=0.12×2.5 µm², phosphorus SIC implant, nominal W_S, and large spacing D. As can be seen from Fig. 16, a reduction of about 56% in W_{SIC} significantly reduces C_{SIC} and leads to about 28% decrease in C_{CB} (from 4.93 fF to 3.56 fF). As a result, f_{MAX} improves by 46 GHz (from 316 GHz to 362 GHz) while f_T decreases by about 9 GHz, which is believed to be due to an increase in R_C[30]. We note that R_B remained approximately constant with W_{SIC} reduction.

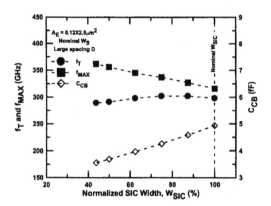

Fig. 16. Simulated C_{CB}, f_T, and f_{MAX} as a function of normalized SIC width (W_{SIC}) for a device with large D.

A significant lateral scaling parameter in SiGe HBTs is the reduction in the emitter width W_E, which can be scaled down to sub-100 nm dimensions using state-of-the-art lithography tools used in CMOS technologies. Simulation results plotted in Fig. 17 show the effect of W_E scaling on R_B, f_T, and f_{MAX} for a device with an emitter length L_E=2.5 μm, phosphorus SIC implant of nominal W_{SIC}, nominal W_S, and large spacing D. Reducing W_E from 0.12 μm to 0.07 μm results in about 33% reduction in R_B (from 43.3 Ω to 29 Ω) caused mainly due to a significant reduction in R_{int} and R_{link}. However, simulation results did not predict a reduction in C_{CB}, as expected from experimental results shown in Fig. 11, which remained about 4.9 fF for all simulated values of W_E. Nevertheless, the decrease in R_B with W_E results in a significant increase of 114 GHz in f_{MAX} (from 314 GHz to 428 GHz) and a moderate decrease in f_T by 11 GHz.

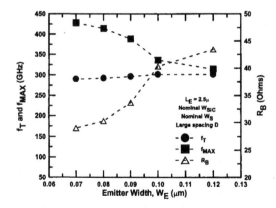

Fig. 17. Simulated R_B, f_T, and f_{MAX} as a function of emitter width (W_E) for a device with large D.

The extrinsic base components R_{spr} and R_{link} can be minimized to further reduce R_B by reducing the spacer width W_S. A narrower spacer in effect reduces the width of the lightly doped base region below the spacer, which reduces R_{spr}, and increases the overlap link area between the heavily doped polysilicon extrinsic base and the intrinsic base, which reduces R_{link}. Simulation results in Fig. 18 demonstrate the effect of normalized W_S (nominal W_S=100%) on R_B, f_T, and f_{MAX} for a device with an emitter area A_E=0.10×2.5 μm^2, phosphorus SIC implant of nominal W_{SIC}, and large spacing D. As can be seen in Fig. 18, R_B decreases significantly with W_S, which is a major advantage of a self-aligned device structure. As W_S is reduced by about 45% below its nominal value, R_B decreases by about 43% (from 40.3 Ω to 23.1 Ω), which results in an increase of 68 GHz in f_{MAX} from 336 GHz to 404 GHz (~20%). On the other hand, when W_S is increased by about 45% above its nominal value, R_B increases slightly by about 18% (from 40.3 Ω to 48.1 Ω), resulting in a small increase of 23 GHz in f_{MAX} (~7%). This indicates that for a large W_S, the extrinsic base resistance component R_{spr} becomes less effective in R_B optimization, in a device with a large spacing D, compared to the other extrinsic base components (e.g. R_{poly}).

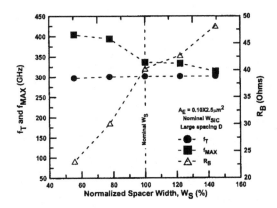

Fig. 18. Simulated R_B, f_T, and f_{MAX} as a function of spacer width (W_S) for a device with large D.

To reduce the extrinsic resistance component R_{poly}, the width of the un-silicided portion of the extrinsic base polysilicon can be minimized by reducing the silicide-to-emitter spacing D. In this case, the current spreads through a smaller portion of un-silicided extrinsic base polysilicon to reach the silicide edge, which effectively reduces R_B and improves f_{MAX}. Such a structure modification of a device with an emitter area A_E=0.12×2.5 μm^2 resulted in about 14% reduction in R_B and a significant increase of 50 GHz in f_{MAX} (from 300 to 350 GHz)[3], which is also predicted by simulation as shown in Fig. 19. Also shown is the effect of W_E and W_S scaling on R_B, f_T, and f_{MAX} for a device with small spacing D. As can be seen from Fig.19, a reduction in W_E from 0.12 μm to 0.10 μm results in a moderate decrease in R_B of about 6% (from 34.5 Ω to 32.5 Ω) and a slight increase of about 27 GHz in f_{MAX} (from 348 GHz to 375 GHz). However, as W_S is reduced by 45% below its nominal value for a device with an emitter area A_E=0.10×2.5 μm^2, R_B decreases significantly by about 55% (from 32.5 Ω to 14.5 Ω). This indicates the effectiveness of W_S reduction in optimizing R_B for a device with small spacing D, where the contribution of the extrinsic resistance component R_{poly} becomes negligible. As a result, f_{MAX} increases by 65 GHz (from 375 GHz to 440 GHz).

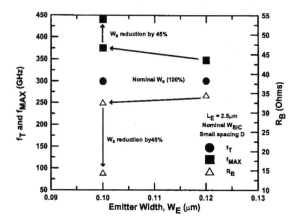

Fig. 19. Simulated R_B, f_T, and f_{MAX} as a function of emitter width (W_E) and spacer width (W_S) for a device with small *D*.

The extrinsic resistance component R_{sil} is determined by the silicide sheet resistance. Cobalt silicide has conventionally been used for SiGe HBT as a low-resistance ohmic contact to the base and collector. However, two high-temperature annealing steps in the range of 500°C-800°C are required to form the lowest resistance phase of cobalt disilicide $CoSi_2$. Such a high thermal budget causes a significant increase in R_B and C_{CB} due to dopants diffusion in the base and collector, which leads to adegradation in both f_T and f_{MAX}. Recent developments in nickel-based silicide implementation in CMOS technology can also be utilized to improve SiGe HBT device performance. The low-resistance mono-silicide phase (e.g. NiSi and NiPtSi) can be formed in one annealing step at low temperatures in the range 400°C-700°C, thus reducing the process thermal budget and dopant diffusion[33,34]. In addition, the lower resistivity and silicon consumption compared to cobalt silicide, is useful in SiGe HBTs to reduce R_B (i.e. R_{sil} component) and improve f_{MAX}.

The advantages of R_{poly} and R_{sil} optimization, however, can be limited by the silicide-to-polysilicon contact resistance. The extrinsic resistance component $R_{sc}=\rho_c/A$ is determined by the silicide contact resitivity (ρ_c) and contact area between the silicide and polysilicon (*A*). The contact resistivity is determined by the doping level at the silicide/polysilicon interface, where high dopant concentration is critical to ensure a low contact resistance. Nickel-based silicides offer lower ρ_c due to lower thermal budget for NiSi and NiPtSi formation[33,34]. Recent results suggested that the contact resistance of NiSi on p-doped substrates is lower than that of $CoSi_2$[35,36], which can be beneficial to reduce R_B (i.e. R_{sc}) and improve f_{MAX} in SiGe HBTs.

4.3. *Extended SiGe HBT Technology Performance Path: Example Case*

Based on the simulation results, few options were selected as an example case, to extend the performance path of SiGe HBT technology. More specifically, as shown in Fig. 20, three options were added by simulation to the performance path in Fig. 13, which include

a) W_E reduction from 0.12 μm to 0.10 μm, b) W_{SIC} reduction by 37% of nominal width, and c) W_S reduction by 45% of nominal width. As a result, as illustrated in Fig. 20, f_{MAX} could be improved to 490 GHz with no significant impact on f_T, which remains about 300 GHz. In addition, further vertical scaling, lateral scaling, and device structure improvements, as described in the previous sections, are expected to enable the evolution of SiGe HBTs towards operational speeds in the THz regime.

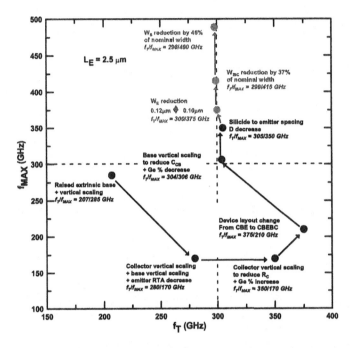

Fig. 19. Extended SiGe HBT technology performance path.

5. Summary

We presented a review of recent developments in SiGe HBT technology that led to a significant improvement in the device performance. High-performance SiGe HBTs operating at speeds approaching 400 GHz have been achieved by vertical scaling, lateral scaling, and device structure innovations enabled by modern CMOS-compatible materials and process techniques. We also presented device simulation results that showed the extendibility of the performance of SiGe HBTs towards half-terahertz and beyond.

6. Acknowledgments

The authors would like to acknowledge partial support of this work by DARPA under SPAWAR contract number N66001-02-C-8014. The authors also would like to thank Joseph Kocis, David Rockwell, Michael Longstreet, Karyn Hurley, and Robert Groves for their support in device process and test.

References

1. J.-S. Rieh, B. Jagannathan, H. Chen, K. Schonenberg, D. Angell, A. Chinthakindi, J. Florkey, F. Golan, D. Greenberg, S. -J. Jeng, M. Khater, F. Pagette, C. Schnabel, P. Smith, A. Stricker, K. Vaed, R. Volant, D. Ahlgren, G. Freeman, K. Stein, and S. Subbanna, SiGe HBTs with cut-off frequency of 350 GHz, in *IEDM Tech. Dig.*, 771-774 (2002).
2. J.-S. Rieh B. Jagannathan, H. Chen K. Schonenberg, S. -J. Jeng, M. Khater, D. Ahlgren, G. Freeman, and S. Subbanna, Performance and design considerations for high speed SiGe HBTs of f_T/f_{MAX} = 375 GHz/210 GHz, in *Proc. of International Conf. on Indium Phosphide and Related Materials*, 374-377 (2003).
3. M. Khater, J. -S. Rieh, T. Adam, A. Chinthakindi, J. Johnson, R. Krishnasamy, M. Meghelli, F. Pagette, D. Sanderson, C. Schnabel, K. Schonenberg, P. Smith, K. Stein, A. Stricker, S. -J. Jeng, D. Ahlgren, and G. Freeman, SiGe HBT technology with f_{MAX}/f_T = 350/300 GHz and gate delay below 3.3 ps, in *IEDM Tech. Dig.*, 247-250 (2004).
4. B. Heinemann, R. Barth, D. Bolze, J. Drews, P. Formanek, T. Grabolla, U. Haak, W. Hoppner, D. Knoll, K. Kopke, B. Kuck, R. Kurps, S. Marschmeyer, H. Richter, H. Rucker, P. Schley, D. Schmidt, W. Winkler, D. Wolansky, H. -E. Wulf, and Y. Yamamoto, A low-parasitic collector construction for high-speed SiGe:C HBTs, in *IEDM Tech. Dig.*, 251-254 (2004).
5. J. Bock, H. Schafer, H. Knapp, K. Aufinger, M. Wurzer, S. Boguth, T. Bottner, R. Stengl, W. Perndl, and T. Meister, 3.3 ps SiGe bipolar technology, in *IEDM Tech. Dig.*, 255-258 (2004).
6. B. Floyd, S. Reynolds, U. Pfeiffer, T. Zwick, T. Beukema, and B. Gaucher, SiGe bipolar transceiver circuits operating at 60 GHz, *IEEE J. Solid-State Circuits* **40**(1), 156-167 (2005)
7. B. Jagannathan, M. Khater, F. Pagette, J. -S. Rieh, D. Angell, H. Chen, J. Florkey, F. Golan, D. Greenberg, R. Groves, S. -J. Jeng, J. Johnson, E. Mengistu, K. Schonenberg, C. Schnabel, P. Smith, A. Stricker, D. Ahlgren, G. Freeman, K. Stein, and S. Subbanna, Self-aligned SiGe NPN transistors with 285 GHz f_{MAX} and 207 GHz f_T in a manufacturable technology, *IEEE Elec. Dev. Lett.* **23**(5), 258-260 (2002).
8. K. Washio, SiGe HBT and BiCMOS technologies, in *IEDM Tech. Dig.*, 113-116 (2003)
9. J. Donkers, T. Vanhoucke, P. Agarwal, R. Hueting, P. Meunier-Beillard, M. Vijayaraghavan, P. Magnee, M. Verheijen, R. De Kort, and J. Slotboom, Metal emitter SiGe:C HBTs, in *IEDM Tech. Dig.*, 243-246 (2004).
10. M. Dahlstrom, X. -M. Fang, D. Lubyshev, M. Urteaga, S. Krishnan, N. Parthasarathy, Y. M. Kim, Y. Wu, J. M. Fastenau, W. K. Liu, M. Rodwell, Wideband DHBTs using a graded carbon-doped InGaAs base, *IEEE Elec. Dev. Lett.* **24**(7), 433-435 (2003).
11. W. Hafez and M. Feng, 0.25 μm Emitter InP SHBTs with f_T = 550 GHz and BV_{CEO} > 2 V, in *IEDM Tech. Dig.*, 549-552 (2004).
12. T. Hussain, Y. Royter, D. Hitko, M. Montes, M. Madhav, I. Milosavljevic, R. Rajavel, S. Thomas, M. Antcliffe, A. Arthur, Y. Boegeman, M. Sokolich, J. Lee, and P. Asbeck, First demonstration of sub-0.25 μm width emitter InP DHBTs with > 400 GHz f_T and > 400 GHz f_{MAX}, in *IEDM Tech. Dig.*, 553-556 (2004).
13. D. Yu, K. Choi, K. Lee, B. Kim, H. Zhu, K. Vargason, J. M. Kuo, P. Pinsukanjana, and Y. Kao, Ultra high-speed 0.25 μm emitter InP-InGaAs SHBTs with f_{MAX} of 687 GHz, in *IEDM Tech. Dig.*, 557-560 (2004).
14. Z. Griffith, M. Dahlstrom, M. Rodwell, X. -M. Fang, D. Lubyshev, Y. Wu, J. Fastenau, and A. Liu, InGaAs-InP DHBTs for increased digital IC bandwidth having 391 GHz f_T and 505 GHz f_{MAX}, *IEEE Elec. Dev. Lett.* **26**(1), 11-13 (2005).
15. Z. Griffith, M. Rodwell, X. -M. Fang, D. Lubyshev, Y. Wu, J. Fastenau, and A. Liu, InGaAs/InP DHBTs with 120 nm collector having simultaneously high f_T and $f_{MAX} \geq$ 450 GHz, *IEEE Elec. Dev. Lett.* **26**(8), 530-532 (2005).

16. W. Snodgrass, B. –R. Wu, W. Hafez, K. –Y. Cheng, and M. Feng, Graded base type-II InP/GaAsSb DHBT with f_T = 475 GHz, *IEEE Elec. Dev. Lett.* **27**(2), 84-86 (2006).

17. N. Parthasarathy, Z. Griffith, C. Kadow, U. Singisetti, M. Rodwell, X. –M. Fang, D. Loubychev, Y. Wu, J. Fastenau, and A. Liu, Collector-pedestal InGaAs/InP DHBTs fabricated in a single-growth, triple-implant process, *IEEE Elec. Dev. Lett.* **27**(5), 313-316 (2006).

18. R. Krithivasan, Y. Lu, J. Cressler, J. –S. Rieh, M. Khater, D. Ahlgren, and G. Freeman, Half-Terahertz operation of SiGe HBTs, *IEEE Elec. Dev. Lett.* **27**(7), 567-569 (2006).

19. S. -J. Jeng, D. Greenberg, M. Longstreet, G. Hueckel, D. Harame, and D. Jadus, Lateral scaling of the self-aligned extrinsic base in SiGe HBTs, in *Proc. BCTM*, 15-18 (1996).

20. S. -J. Jeng, D. Ahlgren, G. Berg, B. Ebersman, G. Freeman, D. Greenberg, J. Malinowski, D. Nguyen-Ngoc, K. Schonenberg, K. Stein, D. Colavito, M. Longstreet, P. Ronsheim, S. Subbanna, and D. Harame, Impact of extrinsic base process on NPN HBT performance and polysilicon resistor in integrated SiGe HBTs, in *Proc. BCTM*, 187-190 (1997).

21. M. Hashim, R. Lever, and, P. Ashburn, Two dimensional simulation of transient enhanced boron out-diffusion from the base of a SiGe HBT due to an extrinsic base implant, in *Proc. BCTM*, 96-99 (1997).

22. J. -S. Rieh, D. Greenberg, M. Khater, K. Schonenberg, S. -J. Jeng, F. Pagette, T. Adam, A. Chinthakindi, J. Florkey, B. Jagannathan, J. Johnson, R. Krishnasamy, D. Sanderson, C. Schnabel, P. Smith, A. Stricker, S. Sweeney, K. Vaed, T. Yanagisawa, D. Ahlgren, K. Stein, and G. Freeman, SiGe HBTs for millimeter-wave applications with simultaneously optimized f_T and f_{MAX} of 300 GHz, in *IEEE RFIC Symp. Dig.*, 395-398 (2004).

23. B. Meyerson, Low-temperature silicon epitaxy by ultrahigh vacuum/chemical vapor deposition, *Appl. Phys. Lett.* **48**(12), 797-799 (1986).

24. B. Meyerson, UHV/CVD growth of Si and Si:Ge alloys: chemistry, physics, and device applications, *Proc. IEEE* **80**(10), 1592-1608 (1992).

25. D. Harame, J. Comfort, J. Cressler, E. Crabbe, J. Sun, B. Meyerson, and T. Tice, Si/SiGe epitaxial-base transistors-part I: materials, physics, and circuits, *IEEE Trans. Elec. Dev.* **42**(3), 455-468 (1995).

26. D. Harame, J. Comfort, J. Cressler, E. Crabbe, J. Sun, B. Meyerson, and T. Tice, Si/SiGe epitaxial-base transistors-part II: process integration and analog applications, *IEEE Trans. Elec. Dev.* **42**(3), 469-482 (1995).

27. L. Lanzerotti, J. Sturm, E. Stach, R. Hull, T. Buyuklimanli, and C. Magee, Suppression of boron outdiffusion in SiGe HBTs by carbon incorporation, in *IEDM Tech. Dig.*, 249-252 (1996).

28. M. Xu, S. Decoutere, A. Sibaja-Hernandez, K. Van Wichelen, L Witters, R. Loo, E. Kunnen, C. Knorr, A. Sadovnikov, and C. Bulucea, Ultra low power SiGe:C HBT for 0.18 μm RF-BiCMOS, in *IEDM Tech. Dig.*, 125-128 (2003).

29. A. Joseph, P. Geiss, X. Liu, J. Johnson, K. Schonenberg, A. Chakravarti, D. ahlgren, and J. Dunn, Emitter resistance improvement in high-performance SiGe HBTs, in *Proc. ISTDM*, 53-54 (2003).

30. A. Stricker, G. Freeman, M. Khater, and J. –S. Rieh, Evaluating and designing the optimal 2D collector profile for a 300 GHz SiGe HBT, *Mat. Sci. Semi. Proc.* **8**, 295-299 (2005).

31. J. -S. Rieh, D. Greenberg, A. Stricker, and G. Freeman, Scaling of SiGe heterjunction bipolar transistors, *Proc. IEEE* **93**(9), 1522-1538 (2005).

32. S. -J. Jeng, B. Jagannathan, J. -S. Rieh, J. Johnson, K Schonenberg, D. Greenberg, A. Stricker, H. Chen, M. Khater, D. Ahlgren, G. Freeman, K. Stein, and S. Subbanna, A 210 GHz f_T SiGe HBT with a non-self-aligned structure, *IEEE Elec. Dev. Lett.* **22**(11), 542-544 (2001).

33. T. Morimoto, T. Ohguro, H. Momose, T. Iinuma, I. Kunishima, K. Suguro, I. Katakabe, H. Nakajima, M. Tsuchiaki, M. Ono, Y. Katsumata, and H. Iwai, Self-aligned nickel-mono-silicide

technology for high-speed deep submicron logic CMOS ULSI, *IEEE Trans. Elec. Dev.* **42**(5), 915-922 (1995).

34. P. Lee, K. Pey, D. Mangelinck, J. Ding, D. Chi, and L. Chan, New salicidation technology with Ni(Pt) alloy for MOSFETs, *IEEE Elec. Dev. Lett.* **22**(12), 568-570 (2001).

35. J. Kittl, A. Lauwers, O. Chamirian, M. Van Dal, A. Akheyar, M. De Potter, R. Lindsay, and K. Maex, Ni- and Co-based silicides for advanced CMOS applications, *Microelec. Eng.* **70**(2-4), 158-165 (2003).

36. J. A. Kittl, A. Lauwers, O. Chamirian, M. Van Dal, A. Akheyar, O. Richard, J. Lisoni, M. De Potter, R. Lindsay, and K. Maex, Silicides for 65 nm CMOS and beyond, in *Mat. Res. Soc. Symp. Proc.* **765**, 267-278 (2003).

International Journal of High Speed Electronics and Systems
Vol. 17, No. 1 (2007) 81–84
© World Scientific Publishing Company

OPTICAL PROPERTIES OF GaInN/GaN MULTI-QUANTUM WELL STRUCTURE AND LIGHT EMITTING DIODE GROWN BY METALORGANIC CHEMICAL VAPOR PHASE EPITAXY

J. Senawiratne, M. Zhu, W. Zhao, Y. Xia, Y. Li, T. Detchprohm, and C. Wetzel

Future Chips Constellaition, Department of Physics, Applied Physics, and Astronomy, Rensselaer, Polytechnic Institute,
Troy, NY 12180, USA
senawj@rpi.edu

Optical properties of green emission $Ga_{0.80}In_{0.20}N$/GaN multi-quantum well and light emitting diode have been investigated by using photoluminescence, cathodoluminescence, electroluminescence, and photoconductivity. The temperature dependent photoluminescence and cathodoluminescence studies show three emission bands including GaInN/GaN quantum well emission centered at 2.38 eV (~ 520 nm). The activation energy of the non-radiative recombination centers was found to be ~ 60 meV. The comparison of photoconductivity with luminescence spectroscopy revealed that optical properties of quantum well layers are strongly affected by the quantum-confined Stark effect.

Keywords: GaN; GaInN; QCSE; photoluminescence; cathodoluminescence; electroluminescence; photoconductivity

1. Introduction

Recent development in GaInN/GaN quantum well based light emitting diodes (LED) have gained much attention in lighting industry due to their potential full color display capability in the entire spectral range including wavelengths in the blue, green, yellow, and red[1]. However, the performance of the GaInN/GaN-based green LEDs is far behind that of the blue LEDs due to the difficulties of growth of indium (In) rich GaInN and limited knowledge of light emission processes, especially in green LEDs. Therefore, understanding the optical and structural properties of these materials is vastly benefited for their further development. This study reports on the optical properties of green emission $Ga_{1-x}In_xN$/GaN multi quantum well (MQW) structure and LED grown by metalorganic chemical vapor phase epitaxy (MOVPE).

2. Experimental

The MQW structure has been grown by MOVPE in an Emcore D-180 spectra GaN rotating disc multiwafer system on (0001) plane of sapphire substrate. The MQW consists of five $Ga_{0.80}In_{0.20}N$/GaN quantum wells of nominal well width of 3 nm separated by barriers of nominal width 11 nm. The $Ga_{0.80}In_{0.20}N$/GaN LED has been grown by embedding five MQWs in a pn-diode on (0001) plane of sapphire substrate. A detailed description of the growth procedure can be found elsewhere[2].

Optical properties were mainly investigated by photoluminescence (PL) and cathodoluminescence (CL) spectroscopy from 5.2 K to 296 K using a modified Gatan MonoCL spectrometer attached to a Jeol 6300 scanning electron microscope (SEM) system. PL experiment was performed using HeCd laser of energy 3.82 eV (325 nm) as the excitation light source. In addition, photoconductivity (PC) measurements were carried out using a SpectraPro model 2300i monochromator and halogen excitation light source, while the signal was collected using low noise phase-matched lock-in technique. The spectral resolution of the optical system is better than 1 nm.

3. Results and Discussion

Figure 1 shows the evaluation of the PL and CL spectra for the $Ga_{0.80}In_{0.20}N$/GaN MQWs over a temperature range from 5.2 K to 296 K. Both PL and CL spectra are dominated by strong MQW emission centered at 2.38 eV (520 nm). The emission arising from the annihilation of GaN excitons at neutral donors (D^0X) was observed at 3.45 eV (~ 360 nm). In addition, at low temperature ($\leq$140 K) GaN donor acceptor pair (DAP) transitions were found to be peaked at 3.2 eV (380 nm), while they were superimposed with LO phonon replicas of GaN of energy ~ 90 meV (735 cm^{-1})[3].

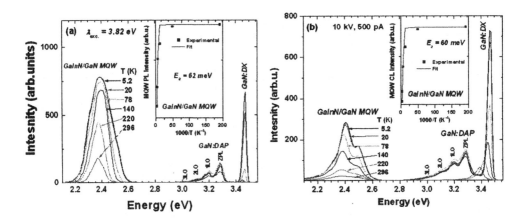

Fig. 1. Temperature dependence (a) PL and (b) CL spectra of $Ga_{0.80}In_{0.20}N$/GaN MQW structure as a function of temperature. The inset of the each figure shows the Arrhenius plot of the PL (and CL) emissions intensity of MQW emission.

The inset of Fig. 1 shows the variation of the intensities of MQW emissions as a function of temperature, and the data was fitted with the following equation[4]:

$$I = I_0[1 + \alpha \, \exp(-E_a/K_B T)] \tag{1}$$

In this model, non-radiative decay is considered to be a thermally activated process which follows a non-radiative lifetime given by Eq. (2)[4].

$$\tau_{nr} = \tau_0 \exp(E_a/K_B T) \tag{2}$$

Where E_a is the activation energy for PL (or CL) quenching, and α is the ratio between radiative lifetime and τ_0. According to the fitting results, the activation energies of PL and CL were found to be 62 meV and 60 meV, respectively. These results are in good agreement with the activation energy of 63 meV for the thermal quenching of the PL intensity of $In_{0.20}Ga_{0.80}0N$ MQWs observed by Teo et al.[5]. The origin of the high activation is most likely associated with the activation energy of the non-radiative recombination centers, such as dislocations.

Furthermore, the optical and electrical analysis of fabricated $Ga_{0.20}In_{0.80}N/GaN$ LED structure were performed using PL, CL, EL, and PC spectroscopic techniques. Fig. 2 shows the comparison of the PL, EL, and PC spectra of $Ga_{0.80}In_{0.20}N/GaN$ LED which consists of bright PL emission centered at 2.29 eV (540 nm), while EL is blueshifted to 2.37 eV (523 nm). It is known that large piezoelectric fields are present in GaInN/GaN structure leading to quantum-confined Stark effect (QCSE)[6], which alters optical properties of the material significantly. The observed blueshift of 80 meV in MQW emission of PL spectrum with respect to that of the EL spectrum clearly reflects the QCSE in the GaInN/GaN LED structure due to the large biaxial strain in the active region. Moreover, the blueshift in EL peak energy with increasing operation current in LED also confirms the existence of the QCSE in the active region of the LED.

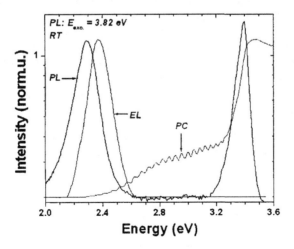

Fig. 2. Comparison of room temperature PL, EL and PC spectra of full processed $Ga_{0.80}In_{0.20}N/GaN$ LED structure.

As shown in Fig. 2, the PC spectrum exhibits a broad absorption band starting at 2.39 eV corresponding to the absorption in the GaInN quantum well layers. In addition, strong absorption in the PC spectrum at 3.44 eV is assigned to the absorption from the GaN layer. In addition, comparison of PL and PC shows that PL of the MQW emission is Stokes-shifted by ~ 100 meV compared to the absorption edge of InGaN quantum well layers determined by the PC experiment. The origin of the Stokes shift may have originated either by QCSE and/or indium composition fluctuation. However, the observed Stokes shift for the PL in PC is quite similar to the PL Stokes shift observed in the EL experiment. In the absence of persistent photoconductivity behavior in the observed structure, it is unlikely that for the observed PL Stokes shift with respect to the PC originated by indium composition fluctuation. Therefore, the observed Stokes shifts are most-likely directly associated with the QCSE, while contributions from indium composition fluctuations in the active region are negligible.

4. Conclusion

We have investigated the optical properties of $Ga_{0.80}In_{0.20}N$/GaN MQW and LED structures grown by MOVPE. From the temperature dependent PL and CL results we found that the activation energy of the non-radiative recombination centers is ~60 meV. Comparing, PC and EL with PL results, we conclude that the QCSE as the dominant mechanism in the Stokes shift in GaInN/GaN MQW emission, while the contributions from indium composition fluctuation are negligible.

References

1. I. Akasaki and H. Amano: High Brightness Light Emitting Diodes, eds. G. B. Stringfellow and M. G. Craford (Academic Press, London, 1997) Semiconductors and Semimetals Vol. **48**, p. 357.
2. C. Wetzel, T. Salagaj, T. Detchprohm, P. Li, J. S. Nelson, GaInN/GaN growth optimization for high-power green light emission diodes, Appl. Phys. Lett. **85**, 866–868 (2004).
3. J. Senawiratne, M. Strassburg, A. Payre, A. Asghar, W. Fernwick, N. Li, I. Ferguson, and N. Dietz, Optical Transitions in MOCVD Grown Cu Doped GaN, Mater. Res. Soc. Symp. Proc. **892**, 1-6 (2006).
4. M. H. Crawford, J. Han, M. A. Banas, S. M. Myers, G. A. Petersen, and J. J. Figiel, Optical spectroscopy of InGaN epilayers in the low indium composition regime, MRS Internet J. Nitride Semicond. Res. 5S1, W11.41 (2000).
5. K. L. Teo, J. S. Colton, P. Y. Yu, E. R. Weber, M. F. Li, W. Liu, K. Uchida, H. Tokunaga, N. Akutsu, and K. Matsumato, An analysis of temperature dependent photoluminescence line shapes in InGaN, App. Phys. Lett. **73**, 1697-1699 (1998).
6. C. Wetzel, T. Takeuchi, H. Amano, I. Akasaki, Piezoelectric Stark-like ladder in GaN/GaInN/GaN heterostructures, Jpn. App. Phys. Lett. **38**, L163-L165 (1999).

International Journal of High Speed Electronics and Systems
Vol. 17, No. 1 (2007) 85–89
© World Scientific Publishing Company

Electrical Comparison of Ta/Ti/Al/Mo/Au and Ti/Al/Mo/Au Ohmic Contacts on Undoped GaN HEMTs Structure with AlN Interlayer

Yunju Sun, and Lester F. Eastman

School of Electrical and Computer Engineering, Cornell University, 426 Phillips Hall,
Ithaca, NY 14853, USA
ys99@cornell.edu

A significant improvement of contact transfer resistance on undoped GaN/AlGaN/AlN (10 Å)/GaN high electron mobility transistor (HEMT) structure was demonstrated using a Ta/Ti/Al/Mo/Au metallization scheme compared to a Ti/Al/Mo/Au metallization scheme. A contact resistance as low as 0.16 ± 0.03 ohm-mm was achieved by rapid thermal annealing of evaporated Ta (125 Å)/Ti (150 Å)/Al (900 Å)/Mo(400 Å)/Au(500 Å) metal contact at 700 °C for 1 min followed by 800 °C for 30 sec in a N_2 ambient. An excellent edge acuity was also demonstrated for the annealed Ta/Ti/Al/Mo/Au ohmic contacts.

Keywords: ohmic contact, GaN, AlGaN/GaN, HEMT

1. Introduction

Recently, a Ti/Al/Mo/Au ohmic metallization system with a pretreatment of $SiCl_4$ plasma in a RIE system demonstrated excellent ohmic electrical performance on both n-GaN and n-type doped AlGaN/GaN surface [1], [2]. Later, for Ti/Al/Mo/Au ohmic metallization system, its electrical performance related to the metal thickness evaporated and annealing condition has been optimized on the Undoped AlGaN/GaN heterostructure [3] without any plasma pretreatment.

We found that, based on the optimized Ti/Al/Mo/Au metallization system, it was more difficult to form ohmic contact with a low contact resistance on an undoped AlGaN/AlN/GaN heterostructure than on an undoped AlGaN/GaN material structure without an AlN interlayer. The AlN interlayer helps to enhance the sheet charge confinement and improve electron mobility at the heterojunction interface [4].

Tantalum has a lower work function (4.25 eV) comparing to Ti (4.33 eV). It has also been shown that Ta can react with N to form TaN [5] and the outdiffusion of nitrogen from AlGaN during annealing results in the formation of an n-type doped layer under contacts. In this work, we present a Ta/Ti/Al/Mo/Au ohmic metallization system on undoped GaN (cap layer)/AlGaN/AlN/GaN HFETs structure with the top four layers are fixed at Ti (15 nm)/Al (90 nm)/Mo (40 nm)/Au (50 nm) as optimized in [3] considering both the electrical performance and edge acuity. At the same time, Ta/Ti/Al/Mo/Au and Ti/Al/Mo/Au metallization schemes were fabricated on the same chip and the results were compared with each other. A significant improvement in the electrical performance using Ta/Ti/Al/Mo/Au on undoped GaN (cap layer)/AlGaN/AlN/GaN compared to

Ti/Al/Mo/Au is demonstrated. The edge acuity of Ta/Ti/Al/Mo/Au metallization system is also examined using scanning electron microscopy (SEM).

2. Experimental results and discussion

A whole set of three different ohmic metal stacks was fabricated on a chip, which size is 13 mm x 13 mm. They are Ta (75 Å)/Ti/Al/Mo/Au, Ta (125 Å)/Ti/Al/Mo/Au, and Ti/Al/Mo/Au respectively. The material structure grown by MOCVD from top to bottom consisted of GaN cap (10 Å, undoped)/ $Al_{0.30}Ga_{0.70}N$ (240 Å, undoped)/ AlN inter-barrier (10 Å)/ SI-GaN buffer layer on a SiC substrate. The rectangular TLM patterns were defined by photolithography. The sample was cleaned using buffered oxide etch (BOE) 30:1 for 30 sec right before metal evaporation. The real spacing between two metal pads after lift-off was varied from 5 μm to 35 μm in six steps. Mesa isolation for the TLM contact test pattern was achieved, after metal lift-off, by inductively coupled plasma (ICP) reactive ion etching system. The chlorine species chemically removes GaN-based material.

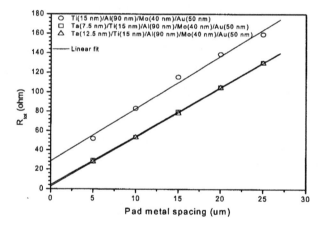

Fig. 1. Total resistance and its linear fit as a function of the pad metal spacing for Ti/Al/Mo/Au, Ta (75 Å)/Ti/Al/Mo/Au, and Ta (125 Å)/Ti/Al/Mo/Au annealed at 700 °C for 1 min followed by 800 °C for 30 sec.

The 13 mm x 13 mm chip was rapid thermal annealed at 700 °C for 1 min followed by 800 °C for 30 sec in a N_2 ambient. The sheet resistance, R_{sh}, (ohm/square) of the bulk semiconductor and contact transfer resistance, r_t, (ohm-mm) were measured by four-point-probe measurement technique. A sheet resistance, R_{sh}, of 500 ohm/square was measured. The total resistance and its linear fit as a function of the pad metal spacing for Ta (75 Å)/Ti/Al/Mo/Au, Ta (125 Å)/Ti/Al/Mo/Au, and Ti/Al/Mo/Au metal stacks is shown in Figure 1, which was measured on one of the TLM patterns for each of the metal stacks. The correlation coefficient, which parameterized the quality of linear-fit, was 0.9997-0.9999 for Ta (75 Å)/Ti/Al/Mo/Au and Ta (125 Å)/Ti/Al/Mo/Au ohmic contacts.

As shown in Figure 1, their total resistance increased linearly with the pad metal spacing with almost no scatter.

Table I summarizes the results of the dc contact transfer resistance of the above three different metal stacks annealed at 700 °C for 1 min followed by 800 °C for 30 sec. As shown in table I, there was a significant improvement in contact resistance from 2.05 ± 0.63 ohm-mm using Ti/Al/Mo/Au to 0.16 ± 0.03 ohm-mm using Ta (125 Å)/Ti/Al/Mo/Au as metal contacts on an undoped GaN (cap layer)/AlGaN/AlN (10 Å) /GaN HEMT structure.

Table 1. A summary of the dc transfer contact resistance of three different metal stacks on sample No.2 annealed at 700 °C for 1 min followed by 800 °C for 30 sec with a measured sheet resistance, R_{sh}, of 500 ohm.

Metallization Scheme	r_t (ohm-mm)	Annealing Condition
Ta 125 Å/Ti/Al/Mo/Au	0.16 ± 0.03	700 °C 1min + 800 °C 30 sec
Ta 75Å/Ti/Al/Mo/Au	0.22 ± 0.02	700 °C 1min + 800 °C 30 sec
Ti/Al/Mo/Au	2.05 ± 0.63	700 °C 1min + 800 °C 30 sec

The change of contact transfer resistance as a function of annealing temperature for Ta-based metallization systems on the same chip was also studied, as shown in Figure 2. They were all first annealed at 700 °C for 1 min and then annealed at 800, 820, and 845 °C respectively for 30 sec. For Ta at a thickness of 125 Å, the contact resistance increased gradually with the annealing temperature from 0.16 ± 0.03 ohm-mm annealed at 800 °C, whereas for Ta at a thickness of 75 Å, the lowest contact resistance of 0.17 ± 0.03 ohm-mm was achieved at an annealing temperature of 845 °C. Both of the Ta/Ti/Al/Mo/Au metal stacks discussed here reached a similar value of dc contact transfer resistance at different annealing temperature, which was the lowest value reported to date on undoped GaN (cap layer)/AlGaN/AlN (10 Å) /GaN HEMT structure.

The edge acuity of Ta/Ti/Al/Mo/Au ohmic contacts was also examined under SEM after they were annealed at a temperature of 700 °C for 1 min followed by 800 °C for 30 sec. As shown in Figure 3, a good edge acuity was demonstrated after the high temperature annealing.

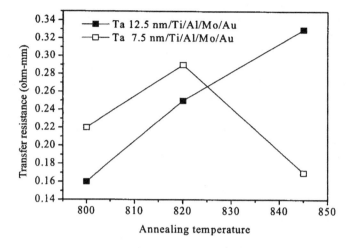

Fig.2. The change of contact resistance as a function of annealing temperature for Ta/Ti/Al/Mo/Au ohmic contacts with Ta at a thickness of 125 Å and 75 Å respectively on sample No.2. They were all first annealed at 700 °C for 1min and then annealed at 800, 820, and 845 °C respectively for 30 sec.

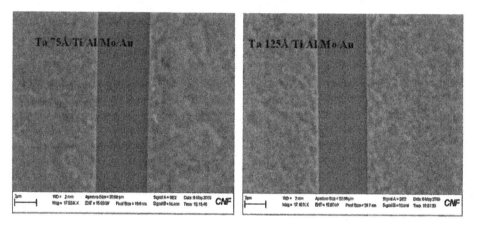

Fig.3. Edge acuity of annealed (700 °C for 1 min + 800 °C for 30 sec) Ta/Ti/Al/Mo/Au ohmic contacts with Ta at a thickness of 75 Å and 125 Å respectively.

3. Conclusions

A successful improvement in ohmic contact resistance using Ta/Ti/Al/Mo/Au metallization scheme was demonstrated compared to the standard Ti/Al/Mo/Au recipe on an undoped GaN (cap layer)/AlGaN/AlN/GaN HEMT structure. A contact resistance as low as 0.16 ± 0.03 ohm-mm, with a sheet resistance of 500 ohm/square, was obtained.

Also, it showed good edge acuity under high temperature annealing. These characteristics make Ta/Ti/Al/Mo/Au a good candidate as an ohmic structure in terms of material and short channel HEMTs design for high frequency performance.

Reference

[1]. V. Kumar, L. Zhou, D. Selvanathan, and I, Adesida, "Thermally-stable low-resistance Ti/Al/Mo/Au multilayer ohmic contacts on n-GaN," *J. Appl. Phys.*, vol. 92, no. 3, pp. 1712-1714, 2002.

[2]. D. Selvanathan, L. Zhou, V. kumar, and I. Adesida, " Low resistance Ti/Al/Mo/Au ohmic contacts for AlGaN/GaN heterostructure field effect transistors," *Phys. Stat. Sol., (a) 194, No. 2, pp. 583-586, 2002.*

[3]. Yunju Sun, Xiaodong Chen, and L. F. Eastman, "Comprehensive study of Ohmic electrical characteristics and optimization of Ti/Al/Mo/Au multilayer ohmics on undoped AlGaN/GaN heterostructure." *J. Appl. Phys.*, (2005), 98, 053701.

[4]. L. Shen, S. Heikman, B. Moran, R. Coffee, N.-Q Zhang, D. Buttari, I. P. Smorchkova, S. Keller, S. P. Denbaars, and U. K. Mishra, "AlGaN/GaN high power microwave HEMTs," IEEE Electron Device Lett., vol. 22, pp. 457-459, Oct 2001.

[5]. B.P. Luther, J. M. DeLucca, S. E. Mohney, and R. F. Karlicek, Jr., *Appl. Phys. Lett.* 71, 3859 (1997).

International Journal of High Speed Electronics and Systems
Vol. 17, No. 1 (2007) 91–95

World Scientific
www.worldscientific.com

ABOVE 2 A/mm DRAIN CURRENT DENSITY OF GaN HEMTS GROWN ON SAPPHIRE

F. Medjdoub[1], J.-F. Carlin[2], M. Gonschorek[2], E. Feltin[2], M.A. Py[2], N. Grandjean[2], and E. Kohn[1], *Member IEEE*

[1]University of Ulm, Albert Einstein Allee 45, 89081 Ulm, Germany,

Email: *farid.medjdoub@uni-ulm.de*

[2]Institute of Quantum Electronics and Photonics, Ecole Polytechnique Fédérale de Lausanne (EPFL),

CH 1015 Lausanne, Switzerland

Keywords: GaN HEMT, power, AlInN

We report on the investigation of an InAlN/GaN HEMT structure, delivering higher sheet carrier density than the commonly used AlGaN/GaN system. We achieved in a reproducible way more than 2 A/mm maximum drain current density for a gate length of 0.25 μm with unpassivated undoped devices realized on sapphire substrates. Small signal measurements yield a $F_T = 31$ GHz and $F_{MAX} = 52$ GHz, which illustrates the capability of these structures to operate at high frequencies. Moreover, the pulsed analysis indicates a more stable surface in the case of AlInN than that of AlGaN, attributed to the lattice matched growth of this barrier with 17 % In content on GaN, avoiding strain piezo polarization in the material.

1. Introduction

GaN is a universal material for a variety of electronic and optoelectronic applications such as light emitters, transistors, and sensors. Due to its wide band gap it is particularly interesting for high-frequency and high-power Field-Effect Transistors devices. However, many difficulties arise from the lattice mismatch and polar surface of the commonly used AlGaN/GaN heterojunction. Even if several improvements have been made in the growth and design of this material system in recent years the commercialization remains still difficult. Recently, the AlInN/GaN heterojunctions were proposed as alternative for power FETs [1].

In this paper we discuss results obtained from a GaN-based High Electron Mobility Transistor (HEMT) with an AlInN barrier. Applying Vegard's law for alloys of InN, AlN and GaN as indicated in Fig. 1, InAlN layers containing 17% Indium can be grown lattice matched on GaN and therefore have no piezoelectric stress component. This rather novel configuration promises high current densities up to 3 A/mm, the 2DEG being only induced by the difference in spontaneous polarization. Moreover, it offers a higher band offset between the GaN channel and the InAlN barrier as compared to AlGaN. However, the materials system is difficult to grow due to the distinctly different growth conditions required for GaN and InAlN and the difficulty to incorporate In and Al in the same lattice without clustering. However, these problems have been widely overcome in first promising investigations [2,3].

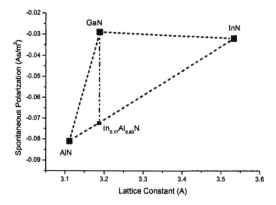

Figure 1: Spontaneous polarization over the lattice constant of InN, AlN, GaN. Lattice matched AlInN is obtained for AlInN with 17% Indium.

2. Growth and processing

Major issues for the growth of AlInN are the different requirements for the growth parameters to achieve good crystalline quality for binaries contained in the ternary alloy, namely InN and AlN. AlN tends to grow polycrystalline for growth temperatures below 1000 °C [4,5]. On the other hand, InN can only be grown at low temperatures and high ammonia partial pressures, whereas high ammonia partial pressures have a strong effect on the AlN growth rate. Here, the studied structures (see Fig. 2) were grown by an AIXTRON MOCVD system on 2 inch diameter (0001) sapphire substrates [6]. They consist of 2 μm thick GaN buffer, 1 nm thick AlN spacer layer and 13 nm thick AlInN barrier layer with 81% Al composition. The thin optimized AlN interlayer is used to reduce alloy disorder scattering and thus to improve the transport characteristics. Room temperature Hall effect measurements yielded a 2DEG sheet charge of 2.6×10^{13} cm^{-2} with a record electron mobility of 1170 cm^2/Vs and a sheet resistivity of 211 $\Omega/\square$ [7].

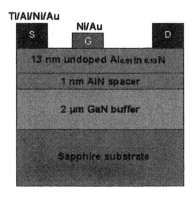

Figure 2: Schematic cross section of the AlInN/GaN HEMT devices

Capacitance-Voltage profiling indicates a 2 DEG carrier density 14 nm below the surface at the AlInN/GaN interface (see Fig. 3), resulting in a sheet charge density in agreement with Hall measurements.

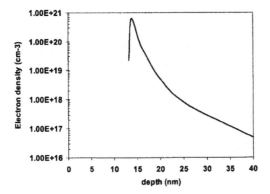

Figure 3: Capacitance-Voltage profiling. Formation of a 2 DEG is indicated at 14 nm below the surface.

FETs have been realized as follows: The devices are mesa isolated by Ar plasma etching. Then, Ti/Al/Ni/Au ohmic contact stacks are alloyed at 890°C in nitrogen atmosphere to provide contact resistances of 0.7 Ω.mm, as measured by linear TLM. The submicron gates ($L_g = 0.25$ μm) are defined with electron beam lithography. Ni/Au is deposited to produce the gate Schottky contacts. The HEMT devices are not yet passivated.

3. Experimental results

The output characteristics of a 0.25 μm gate length device with 25 μm gate width are shown in Fig. 4. The maximum drain current density at +2 V V_{GS} exceeds 2 A/mm in a reproducible way (2.3 A/mm for the presented device), which is to our knowledge beyond the highest drain current density of any AlGaN/GaN HEMT structure, especially when fabricated on sapphire. It reflects the high sheet carrier density in this novel heterostructure. The transconductance peaks at 265 mS/mm (see Fig. 5) with 10 V applied to the drain and -5 V to the gate. The breakdown voltage at pinch-off is estimated to about 40 V for a device with a drain source spacing of 2.5 μm and is thought to be linked to residual buffer layer leakage.

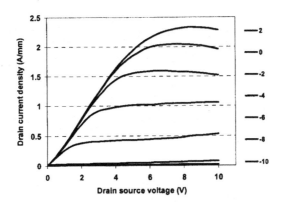

Figure 4: DC output characteristics of an $Al_{0.81}In_{0.19}N$/GaN HEMT 0.25×25 μm². V_{GS} swept from -10 V to 2 V by step of 2V

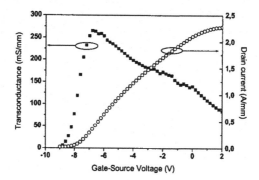

Figure 5: Transfer characteristics of an $Al_{0.81}In_{0.19}N/GaN$ HEMT 0.25×25 μm^2 at V_{DS} = 10V.

Pulse experiments were performed in a routine as used to assess the stability of AlGaN/GaN devices (described with more details in [8]) with 500 ns pulses (see Fig. 6). All quiescent bias points (V_{DS0}, V_{GS0}) are chosen in order to simultaneously eliminate the thermal effect (cold polarization) and to reveal the gate and drain lag effects: (V_{DS0} = 0 V, V_{GS0} = 0 V), (V_{DS0} = 0 V, V_{GS0} = pinch-off voltage) and (V_{DS0} = 10 V, V_{GS0} = pinch-off voltage). The first quiescent bias point is used as the reference to compare with the other bias conditions. A drop of 13 % (gate lag) and 38 % (drain lag) regarding the maximum drain current density is observed at V_{DS} = 10 V. Compared to typical unpassivated AlGaN/GaN structures on sapphire, those changes are much lower, which indicate a more stable surface in the case of AlInN.

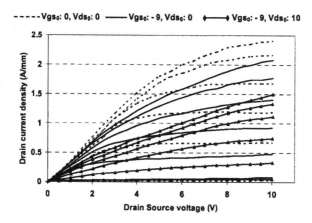

Figure 6: Pulsed I_D-V_{DS} characteristics of an $Al_{0.81}In_{0.19}N/GaN$ HEMT 0.25×50 μm^2 with three different quiescent bias points. V_{GS} swept from -9 to 1 V by step of -2 V.

The S-parameters of 0.25×50 μm^2 HEMTs were measured between 0.5 and 40 GHz. We achieved cut-off and maximum oscillation frequencies F_t = 31 GHz and F_{max} = 52 GHz (shown in figure 7) extrapolated from the current gain H_{21} and the maximum available gain (MAG) at V_{GS} = -6.5V and V_{DS} = 10V respectively. These results show the capability of this structure to operate at high frequency. It reflects the high material quality already obtained and the maturity of the processing technology. We observed a good F_{MAX} / F_T ratio, which is may be directly correlated to the device power performances.

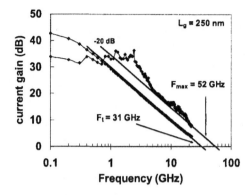

Figure 7: Current gain cut-off frequency ($|H_{21}|$) and maximum oscillation frequency (extrapolated from the Maximum Available Gain) for a 0.25×50 µm² $Al_{0.81}In_{0.19}N/GaN$ HEMT

4. Conclusion

We have investigated the potential of AlInN/GaN structures for high power applications. We obtained more than 2 A/mm open channel current which corresponds to the highest maximum drain current density delivered by any GaN FET. We have also revealed the capabilities of these structures to operate at high frequencies. Pulsed measurements reveal a more stable surface in the case of AlInN than that of AlGaN. However, the breakdown voltage of these devices is still limited mainly because of a relatively low GaN buffer resistivity. The improvement of the GaN buffer resistivity associated with the utilization of a field plate technology should strongly improve the three-terminal breakdown voltage of these structures. Power measurements will be carried out on these devices.

Acknowledgement

This study is supported by the European Union under contract No. 6903

References

1. J. Kuzmik, "Power Electronics on InAlN/(In)GaN: Prospect for a Record Performance," *IEEE Electron Device Lett.*, Vol. 22, p. 510, 2001
2. M. Neuburger, *et al.* "Unstrained InAlN/GaN FET," *Int. J. High Speed Electron. Syst.*, Vol. 14, p. 785, 2004
3. F. Medjdoub, *et al.* "Small signal characteristics of AlInN/GaN HEMTs," *Electronics Lett.*, Vol. 42, p. 779, 2006
4. A. Watanabe, *et al.* "The growth of single crystalline GaN on a Si substrate using AlN as an intermediate layer," *J. Cryst. Growth* 128, 391, 1993
5. A. Krost, *et al.* "GaN-based epitaxy on silicon: stress measurements," *Phys. Status Solidi A* 200, 26 (2003).
6. J.-F. Carlin, *et al.* "High-quality AlInN for high index contrast Bragg mirrors lattice matched to GaN," *Applied Physics Letters*, Vol. 83, p. 668, 2003
7. M. Gonschorek, *et al.* "High electron mobility lattice-matched AlInN/GaN Field-Effect Transistor heterostructures" *Submitted to Applied Physics Letters*
8. C. Gaquière, *et al.* "Pulsed bias/pulsed RF characterization measurement system of FET at constant intrinsic voltages," *Microwave and Optical Technology Lett.*, Vol. 20, p 348, 1999

International Journal of High Speed Electronics and Systems
Vol. 17, No. 1 (2007) 97–101
© World Scientific Publishing Company

FOCUSED THERMAL BEAM DIRECT PATTERNING ON INGAN DURING MOLECULAR BEAM EPITAXY

XIAODONG CHEN, WILLIAM J. SCHAFF, and LESTER F. EASTMAN

School of Electrical and Computer Engineering, Cornell University, 426 Phillips Hall,
Ithaca, NY 14853, USA
xc32@cornell.edu

During InGaN Molecular Beam Epitaxy (MBE) growth, the material surface is exposed to a small diameter pulse laser beam that is controlled by scanning mirrors. Local heating effects are observed at the points of exposure. The materials are characterized by Wavelength Dispersive Spectroscopy (WDS), Scanning Electron Microscopy (SEM), and Photoluminescence (PL). Indium mole fraction of materials is reduced where laser exposure takes place. The effect of local thermal heating appears to enhance surface diffusion while not causing ablation or evaporation under the conditions studied. PL efficiency is significantly increased by focused thermal beam exposure. Laser written regions have 7 times higher PL intensity compared to non-written areas, which might be due to surface texturing that causes higher extraction efficiency.

Keywords: Direct patterning; InGaN; Molecular beam epitaxy.

1. Introduction

Direct-write technologies are the most recent and novel approaches to the fabrication of electronic devices whose sizes range from the meso- to the nanoscale[1]. Thanks to material advances and processing technique improvement, many different direct-write technologies have been developed, such as plasma spray, laser particle guidance, laser chemical vapor deposition (LCVD), micropen, and ink jet printing, etc[2-6]. These new direct-write technologies have showed a wide range of applications in material processing, electronic, chemical and biological sensors, integrated power sources, 3-D artificial tissue engineering.

Focused thermal beam is one of important transfer processing tools and has been used in many direct-write technologies. At Naval Research Laboratory, patterns of viable bacteria have been successfully direct-written on various substrates with a laser-based technique. In low temperature direct-write processing, laser sintering has been used to enhance particle-particle bonding and electronic properties[1].

So far, there has been no any direct-write technique advanced in III-nitrides. In this work, we report the first successful direct-write patterning of InGaN during molecular beam epitaxy (MBE) using in-situ focused thermal beam. It is shown, that this new technique can achieve lateral InGaN mole fraction variations and significantly improve material optical properties. It is envisioned, that this new technique has wide spread applications in opto-electronics.

2. Experimental details

The direct writing laser system we used is a commercial IPG photonics pulsed fiber laser integrated with a scan head and a beam expander, which has 10 W average power, and up to 0.5 mJ pulse energy. Inside the scan head, there are several mirrors coupled to servos for controlling the laser beam focus location. Commercial laser writing control software, WinLase Professional, was used to design the pattern. We chose 1063 nm laser wavelength for this study, and 10 mm diameter laser beam, which provide a 50 μm size laser spot on the wafer surface.

All InGaN samples in this work were grown by a Varian GEN-II gas-source MBE system. The substrates used were (0001) sapphire wafers with backside sputtered Ti/W alloy at about 1 μm for efficient radiant heat absorption during the growth. The buffer layers used is a 250 nm AlN layer followed by a 1 μm GaN layer grown at high temperature. The strong and streaky reflection high-energy electron diffraction (RHEED) pattern during the AlN and GaN growth indicates good crystalline quality of the buffer layers. GaN buffer was identified to be Ga-face and was used for growing a GaN/AlGaN high electron-mobility transistor structure under the identical conditions. Then the InGaN epilayers were grown at around 530 °C with a growth rate near 0.5 μm/hr. Indium mole fraction is aimed to be 0.8. The surface of epilayer was exposed to the directed laser beam during the growth of $In_xGa_{1-x}N$. The materials were characterized by Wavelength Dispersive Spectroscopy (WDS), Scanning Electron Microscopy (SEM) and Photoluminescence (PL) after patterning.

3. Results and discussion

Figure 1 displays the SEM images of the laser-written patterns during the growth of 80 nm $In_{0.8}Ga_{0.2}N$ layer on the top of 540 nm un-written $In_{0.8}Ga_{0.2}N$ layer. It shows that the

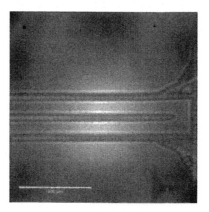

Fig. 1. SEM image of the pattern lines written by the laser beam.

lines are pretty sharp even they have been repeatedly written for 771 passes over 11 minutes. Individual spot can be seen clearly, and the line spacing is approximately 70μm.

In mole fraction was measured quantitatively by WDS, as seen in Fig. 2 (a). It is found that In mole fraction of written materials is reduced from 0.85 where is adjacent to laser exposure, to 0.75 where exposure takes place, while it is 0.81 away from exposed regions. It should be noted, that in our WDS analysis, the penetration depth of the beam is much deeper than the region where the surface diffusion takes place. The absolute number we get here is the average In mole fraction along the penetrating direction. It is likely that the real surface In composition variation will be much bigger than 10%.

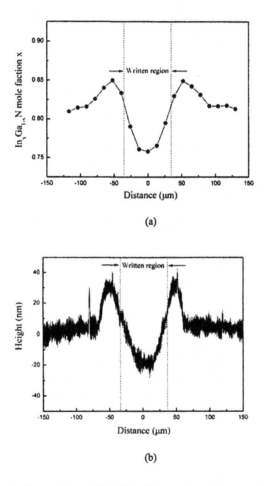

(a)

(b)

Fig. 2. Indium mole fraction variation and height variation across the laser written region. (a) Indium mole fraction variation; (b) Height variation.

Fig. 2 (b) shows the height variation where laser writing has occurred. When we integrate the curve, it is found that the extra materials piling up in the edge regions has the same amount of materials that has moved out of the exposed region. The laser exposure does not appear to evaporate material away from the surface. The wafer is undergoing local heating effects under the laser beam, causing increased surface diffusion of In towards no-illuminated, cooler region.

The photoluminescence (PL) intensity was compared from on and off laser exposed locations, as shown in Fig. 3. It is observed that PL efficiency increases by a factor of 7 from exposed regions to non-written areas. The multi-line scans further confirm this effect, i.e. the PL intensity has a dramatic improvement in the laser written regions. A 2 dimensional growth may have minor variation in efficiency across the wafer, but obtaining such a large increase is completely out of its control. As comparison, PL intensity was measured through the backside of the wafer, but no efficiency improvement was found. That indicates the dominant effect for this increase might not be due to high temperature annealing, but likely comes from surface morphology modification in the written regions. The special texturing was created in the laser-exposed areas that are more effective at extracting light compared to a flat surface.

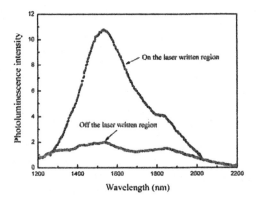

Fig. 3. Room temperature Photoluminescence (PL) intensity compared from on and off laser exposed regions.

4. Conclusion

In summary, in-situ direct patterning of InGaN during MBE growth was first demonstrated. Local heating effects are observed where the laser beam is incident. Focused thermal beam can achieve lateral InGaN mole fractions variations. The effect of local thermal heating appears to enhance Indium surface diffusion while not causing material evaporation under the conditions studied. In addition, PL efficiency is significantly increased by laser beam exposure. This enormous enhancement is believed to be due to surface texturing causing higher extraction efficiency.

Acknowledgments

This work was supported by AFOSR through Georgia Tech MURI subcontract E-21-6T4-G4. And it made use of the Cornell Center for Materials Research Shared Experimental Facilities, supported through the National Science Foundation Materials Research Science and Engineering Centers Program DMR-0079992.

References

1. A. Pique and D. B. Chrisey, eds, *Direct-write technologies for rapid prototyping applications: sensors, electronics and integrated power source* (Academic Press, CA, 2002).
2. H. Esrom, J. Y. Zhang, U. Kogelschatz and A. J. Pedraza, New approach of a laser-induced forward transfer for deposition of patterned thin metal films, *Appl. Surf. Sci.* **86**, 202-207 (1995).
3. M. J. Renn, R. Pastel and H. J. Lewandowske, Laser guidance and trapping of mesoscale particles in hollow-core optical fibers, *Phys. Rev. Lett.* **82**(7), 1574-1577 (1999).
4. M. K. Herndon, R. T. Collins, R. E. Hollingsworth, P. R. Larson and M. B. Johnson, Near-field scanning optical nanolithography using amorphous silicon photoresists, *Appl. Phys. Lett.* **74**(1), 141-143 (1999).
5. B. H. King, D. Dimos, P. Yang and S. L. Morissette, Direct-Write fabrication of integrated, multilayer ceramic components, *J. Electroceram.* **3**(2), 173-178 (1999).
6. D. B. Chrisey, A. Pique, J. F. Gerald, R. C. Y. Auyeung, R. A. McGill, H. D. Wu and M. Duignan, New approach to laser direct writing active and passive mesoscopic circuit elements, *Appl. Surf. Sci.* **154-155**, 593-600 (2000).

Section II.
Terahertz and Millimeter Wave Devices

International Journal of High Speed Electronics and Systems
Vol. 17, No. 1 (2007) 105–110
© World Scientific Publishing Company

TEMPERATURE-DEPENDENT MICROWAVE PERFORMANCE OF SB-HETEROSTRUCTURE BACKWARD DIODES FOR MILLIMETER-WAVE DETECTION

N. SU, Z. ZHANG, and P. FAY[*]

Department of Electrical Engineering
University of Notre Dame, Notre Dame, IN 46556, USA
*pfay@nd.edu

H. P. MOYER, R. D. RAJAVEL, and J. SCHULMAN

HRL Laboratories LLC
3011 Malibu Canyon Rd, Malibu, CA 90265, USA

The temperature dependence of heterostructure backward diodes based on the InAs/AlGaSb/GaSb material system for millimeter-wave detection has been investigated experimentally. Measured dc curvatures of 36 V^{-1} at 298 K and 74 V^{-1} at 4.2 K have been obtained. Variable-temperature on-wafer s-parameters to 110 GHz reveal that the junction capacitance of a typical 2×2 μm^2 area device decreases from 18 fF at 298 K to 11 fF at 77 K, while the junction resistance decreases from 13.9 kΩ to 10.2 kΩ. Directly measured voltage sensitivities at 50 GHz of 3650 V/W and 7190 V/W were obtained at 298 K and 4.2 K, respectively, consistent with the expected value from measured dc curvature. A 1 dB compression point of 18.5 μW and 7.2 μW at 298 K and 77 K, respectively, was measured. A physical model based on self-consistent Poisson-Schrödinger equation solutions was obtained to explain the experimental observations, and suggests the ways to further improve the device performance.

Keywords: backward diode; Sb-heterostructure; tunnel diode; millimeter-wave detection and imaging.

1. Introduction

High-sensitivity detectors at millimeter-wave frequencies are needed for advanced passive imaging arrays and radiometers for security, avionics, and scientific applications. For imaging arrays, direct detection using zero-bias diodes results in significant system-level simplification and reduced generation of 1/f noise. Zero-bias detection using heterostructure backward diodes based on the InAs/AlSb/GaSb material system has been demonstrated, with excellent sensitivity and bandwidth through W-band and above. [1-4]

Investigation of the effect of temperature on detector performance is of importance not only to evaluate the device's robustness against changes in ambient temperature, but also provides valuable information about the underlying device physics. We report here the first experimental investigation of temperature-dependent microwave performance of antimonide-based heterostructure backward diode detectors. To explain the observed dependence of device performance on temperature, a theoretical analysis using self-consistent solution of the Poisson and Schrödinger equations was performed and a physical model was obtained.

2.   Device Structure and Fabrication

The energy band diagram for the device heterostructure as calculated from a self-consistent solution to the Poisson and Schrödinger equations is shown in Fig. 1. The type II band gap alignment in this heterostructure results in an asymmetry in the tunneling current with different bias directions, which produces square-law rectification at zero bias. The device structure shown in Fig. 1 has demonstrated a record-high curvature and room-temperature sensitivity for zero-bias operation.[3] The fabrication of the Sb-based backward diode includes mesa etching, dielectric passivation, via etching, contact metallization and liftoff; details of the device processing have been reported previously.[5]

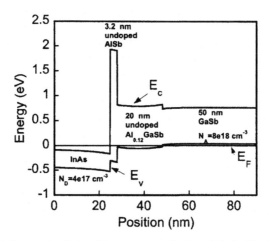

Fig. 1. Calculated band diagram for Sb-based heterostructure backward diode. Key layer compositions and thicknesses are also indicated.

3.   Temperature-dependent Device Performance

The temperature-dependent dc current-voltage characteristics were measured from 4.2 K to 333 K. The curvature, $\gamma=(\partial^2 I/\partial V^2)/(\partial I/\partial V)$, and the junction resistance, $R_j=1/(\partial I/\partial V)$, were calculated from the measured I-V curves and summarized in Fig. 2. The curvature is 36 V^{-1} at room temperature and is approximately constant down to 150 K, below which it rises to 74 V^{-1} at 4.2 K. The junction resistance depends only weakly on temperature as expected for the tunneling-dominated current flow, with R_j changing from 13.9 kΩ at 298 K to 10.2 kΩ at 4.2 K. Therefore, the temperature dependence of γ arises primarily from an increase in the $\partial^2 I/\partial V^2$ term in γ, rather than a decrease in the $\partial I/\partial V$ term.

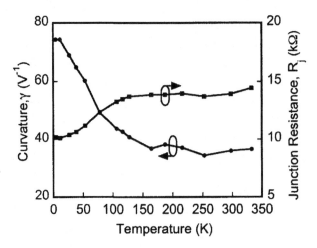

Fig. 2. Measured dc curvature γ and junction resistance R_j vs. temperature.

The detector s-parameters were also measured at different temperatures on-wafer from 1 to 110 GHz. Fig. 3(a) shows the measured and modeled s-parameters obtained by non-linear least square fitting to the equivalent circuit model. The junction resistance, R_j, junction capacitance, C_j, series resistance, R_s, pad inductance, L_p, and pad capacitance, C_p, have been extracted as a function of temperature. For a typical device with active area of 2×2 μm², C_p, L_p, and R_s were found to be independent of temperature (C_p=12 fF, L_p=65 pH, R_s=11 Ω), while both C_j and R_j were found to decrease with decreasing temperature. The junction capacitance is 18 fF at room temperature, and decreases to 11 fF at 77 K as shown in Fig. 3(b). The zero-bias capacitance was also measured at 1 MHz using an HP4280A CV meter, in close agreement with those extracted from the measured s-parameters. The intrinsic cut-off frequency of the detector, defined as $f_c=1/(2\pi C_j R_s)$, is enhanced from 740 GHz at 298 K to 1.1 THz at 77 K, a 48% increase.

The voltage sensitivity of the detector was assessed at 50 GHz (frequency limited by the test fixture) as a function of temperature. The device was wire bonded to an RF test fixture and was driven by an RF source through a bias tee. The detector voltage was measured at the dc arm of the bias tee. Fig. 4(a) shows the measured sensitivity after de-embedding to remove the effects of cable and fixture losses for a zero-biased detector at different temperatures. Measured voltage sensitivities, $β_v$, of 3650 V/W and 7190 V/W were obtained at 298 K and 4.2 K, respectively with incident RF power of 2.5 μW. The increased sensitivity with deceasing temperature is consistent with that expected from the dc curvature by $β_v=2Z_s γ$. The compression performance of the detectors has also been assessed. Fig. 4(b) shows that the zero-bias sensitivity for a typical 2x2 μm² device at 50 GHz is compressed by 1 dB for an incident RF power of 18.5 μW and 7.2 μW at 298 K

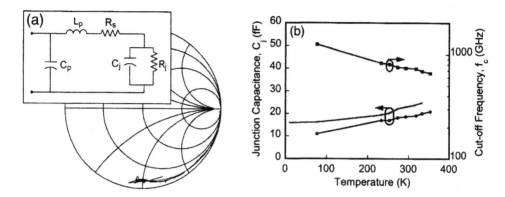

Fig.3. a) Measured and modeled s-parameters from 1-110 GHz for a 2×2 μm^2 area device. Inset is the equivalent circuit model used in the modeling. b) Measured junction capacitance and calculated cut-off frequency vs. temperature. Data points (●) and (■) represent the extracted C_j and calculated f_c based on measured s-parameters. Data points (o) are taken from zero-bias C-V measurement.

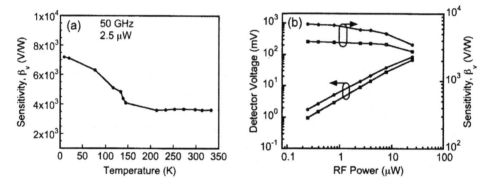

Fig. 4. a) Measured voltage sensitivity vs. temperature at 50 GHz with incident RF power of 2.5 μW. b) Measured detector voltage and sensitivity vs. incident RF power at 50 GHz. Data points (●) and (■) are taken at 77 K and 298 K, respectively.

and 77 K, respectively. The 1 dB compression point decreases with decreasing temperatures, which is expected because the dc curvature has greater variation with RF-induced self-bias at lower temperatures.

4. Theoretical Analysis and Discussions

A theoretical analysis based on self-consistent solution to Poisson and Schrödinger equations was performed. Fig. 5(a) shows the calculated band diagram and Fig. 5(b) shows the charge distribution in the device as a function of lattice temperature. The change of the capacitance with temperature in this calculation is in reasonable agreement with the measured zero-bias capacitance; a 44% decrease in C_j is calculated from 300 K to 77K, compared to a 35% decrease in the measured capacitance. As can be seen in Fig. 5(b), the decreased junction capacitance with decreasing temperature arises from the increased separation of charges due to a combination of the dependence of the band gap and occupation statistics on temperature. In particular, the carrier density in the p-GaSb layer has stronger temperature dependence than that in the n-InAs cathode due to the higher density of states in GaSb. As a result, the decreasing temperature more heavily favors the backward tunnel current compared to the forward current, resulting in an increasingly nonlinear current-voltage characteristic at low temperatures. As suggested in Fig. 5, this nonlinearity of current flow at zero bias is enhanced more rapidly at temperature below 150 K, which is consistent with the observed dependence of curvature on temperature.

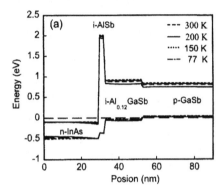 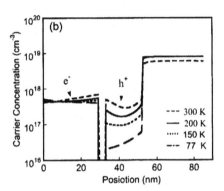

Fig 5. Band diagrams (a) and carrier spatial distribution (b) of the Sb-based heterostructure at several temperatures. Increased separation of charges at lower device temperatures leads to the observed increases in device curvature and decreased device capacitance.

5. Conclusion

The temperature-dependent dc and millimeter-wave performance of InAs/GaAlSb/Sb heterostructure backward diodes have been investigated under zero-bias operation. Junction capacitance and junction resistance are found to decrease with decreasing temperature, while the curvature and directly-measured voltage sensitivity increase with decreasing temperature. To explain these observations, a physical model has been obtained by solving the Poisson-Schrödinger equations as a function of temperature. The theoretical analysis suggests that further increases in room-temperature curvature may be obtained through tailoring of the carrier concentration profile within the device, as well as the possibility of further desensitizing the device to ambient temperature.

6. Acknowledgements

The authors would like to thank Dr. A. Seabaugh for the use of 8510XF network analyzer for the s-parameter measurement, and Dr. A. Orlov for his technical assistance in the low-temperature measurement. This work was funded by the National Science Foundation, ECS-0506950 and IIS-0610169.

References

1. J. N. Schulman, V. Kolinko, M. Morgan, C. Martin, J. Lovberg, S. Thomas, III, J. Zinck, and Y. K. Boegeman, "W-band direct detection circuit performance with Sb-heterostructure diodes," *IEEE Microwave and Wireless Components Lett.*, vol. 14, no. 7, pp. 316-318, 2004.
2. P. Fay, J. N. Schulman, S. Thomas, III, D. H. Chow, Y. K. Boegeman, and K. S. Holabird, "High-performance antimonide-based heterostructure backward diodes for millimeter-wave detection," *IEEE Electron Device Lett.*, vol. 23, no. 10, pp. 585-587, 2002.
3. R. G. Meyers, P. Fay, J. N. Schulman, S. Thomas, III, D. H. Chow, J. Zinck, Y. K. Boegeman, and P. Deelman, "Bias and temperature dependance of Sb-based heterostructure millimeter-wave detectors with improved sensitivity," *IEEE Electron Device Lett.*, vol. 25, no.1, pp. 4-6, 2004.
4. J. N. Schulman, K. S. Holabird, D. H. Chow, H. L. Dunlap, S. Thomas, and E. T. Croke, "Temperature dependence of Sb-heterostructure millimeter-wave diodes," *Electronics Lett.*, vol. 38, no. 2, pp. 94-95, 2002.
5. J. N. Schulman, E. T. Croke, D. H. Chow, H. L. Dunlap, K. S. Holabird, M. A. Morgan, and S. Weinreb, "Quantum tunneling Sb-heterostructure millimeter-wave diodes," in *IEDM Tech. Dig.*, pp. 765-767, 2001.

International Journal of High Speed Electronics and Systems
Vol. 17, No. 1 (2007) 111–114
© World Scientific Publishing Company

A MIXED-SIGNAL ROW/COLUMN ARCHITECTURE FOR VERY LARGE MONOLITHIC mm-WAVE PHASED ARRAYS

CORRADO CARTA[§], MUNKYO SEO and MARK RODWELL

Department of Electrical and Computer Engineering,
University of California at Santa Barbara,
CA 93106-9560, United States of America
[§] *carta@ece.ucsb.edu*

The range of mm-wave radio communications is severely constrained by high losses arising from the short wavelength and from atmospheric attenuation. Large phased arrays can overcome these limitations, but it is very difficult to realize them using present monolithic beamsteering IC architectures. We propose an alternative architecture for large monolithic phased arrays. The beam is steered in altitude and in azimuth by separately imposing vertical and horizontal phase gradients. This choice reduces IC complexity, making large arrays feasible. Since extensive digital processing provides robust amplitude control and reduces die area, the LOs are processed as digital signals. Being very sensitive to compression, the IF signals are processed as analog signals and distributed by means of synthetic transmission-line buses. With careful frequency planning, this mixed-signal approach can allow large phased arrays to operate at frequencies much higher than those achievable with pure analog design.

Keywords: Phased array; Mixed-signal design; mm-Wave.

1. Introduction

The large bandwidth available for mm-wave radio communications allows data rates in the $1 - 10$ Gbit/s range. The usable link range of these channels is severely constrained by the short wavelength and the atmospheric attenuation. The channel loss can be expressed as

$$\frac{P_{rx}}{P_{tx}} = \frac{D_t D_r}{16\pi^2} \left(\frac{\lambda}{R}\right)^2 e^{-\alpha R} \tag{1}$$

where $P_{rx,tx}$ and $D_{rx,tx}$ are power and antenna directivities at receiver and transmitter sides, R is the link range. Apparently, large antenna directivities can enable large ranges: if N-element transmit and receive arrays are used, then $D \sim \pi N$, $P_{tx} \propto N$ and $P_{rx} \propto N^3$. Using 32×32 arrays, link SNR is increased ~ 90 dB, permitting e.g. 10 Gbit/s *mobile* communication over a long ~ 1 km range, even in heavy rain[1]. Two issues arise form the use of arrays. First, unless antennas are fixed, such a high directivity requires electronic beam steering. Second, the element pitch must be smaller than $\lambda/2$, and this constrains maximal area per element in a bi-dimensional array. This area must accommodate beam steering, LNA, PA and

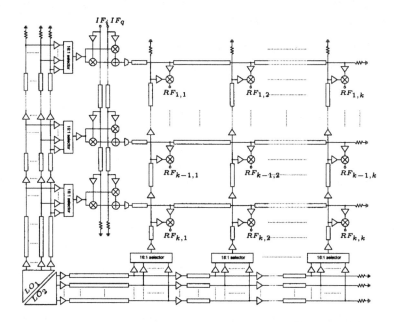

Fig. 1. System architecture of a large monolithic phased array.

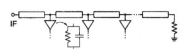

Fig. 2. The synthetic t-line IF bus.

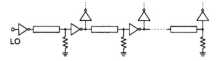

Fig. 3. LOs are processed as digital signals.

T/R switch: as frequency increases, integration becomes necessary. Modern and non-expensive Si technologies offer performances sufficient for the $30 - 100$ GHz range and are, thus, suitable for the integration of monolithic mm-wave arrays.

While the integration of large arrays is attractive, it is very difficult to realize them with present beamsteering IC architectures[2,3]. In this paper, we propose an alternative architecture, which bases its operation on steering the beam in altitude and azimuth by separately imposing vertical and horizontal phase gradients. This simple assumption enables a neat separation of IF and LO distribution lines, and allows improvements in both array size and maximal operation frequency.

2. System Architecture and Design

The system architecture we propose for large monolithic arrays is illustrated in Fig. 1 with a $k \times k$ double-upconversion transmitter. Making use of two LO signals, the input I-Q IF signals are first multiplied by a vertical phase gradient, then an independent horizontal gradient is applied and summed by means of a second up-

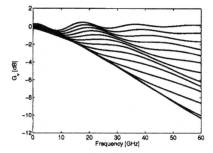
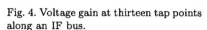

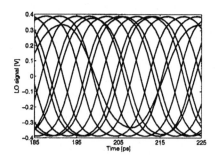

Fig. 4. Voltage gain at thirteen tap points along an IF bus.

Fig. 5. Time domain simulation of a fifteen tap-point 30 GHz LO bus.

conversion mixing. This allows the IC to steer the beam in altitude and azimuth, with minimal wiring complexity. In a N-element array, the number of IF buses and phase selectors grows as $2\sqrt{N}$, not as N. IC complexity is reduced, making large arrays feasible. In terms of maximal array dimension, the fundamental limitation of this architecture is in the maximum number of elements per row and column that can be reached with IF and LO buses without excessive signal degradation.

IF synthetic transmission-line bus — The IF signals are analog and must be processed avoiding compression. We propose to distribute the IF signals along transmission-line buses, which absorb their periodic gate loading capacitance into synthetic transmission-lines. As shown in Fig. 2, the design is similar to a distributed amplifier and avoids area-inefficient reactively-matched subcircuits.

Fig. 4 shows the simulation of an IF bus. The transmission lines are modeled as microstrips in a standard six-metal RF BiCMOS technology (0.18 µm); the line is loaded every 300 µm with the input of an active upconversion mixer in the same technology. The application will set the maximum number of tap points. For example, in applications which can tolerate 2 dB of attenuation, more than 10 tap points could be used at 15 GHz. Depending on the frequency plan, this relatively low IF frequency would not necessarily limit the output RF frequency: a 45 GHz RF would be possible with a first LO at 15 GHz and a second at 30 GHz. The baseband or first IF signal would travel on a low speed IF bus, the second IF signal on a 15 GHz IF bus, in order to be dstributed and mixed with the 30 GHz second LO. This approach, while relaxing significantly the performance requirements for the IF bus, requires a different design strategy for the LO buses, which need to operate at much higher frequencies with the same number of tap points.

LO digital bus — For relatively low frequency applications, or for small-size arrays, the synthetic transmission-line approach can be considered for the LO bus as well: it offers the advantage of being entirely passive. On the other hand, the synthetic transmission line is linear, and common microwave mixers do not need the LO signal to be linear. We propose to trade linearity for maximal frequency of operation by processing LO signals as digital signals, using ECL selectors, transmission-line drivers and repeaters. Fig. 3 shows the schematic of the LO bus: digital inverters are designed to drive terminated transmission-lines; at each tap point the signal is restored. The loading gates can be either mixer LO ports or selector inputs, all ECL compatible. While saturating the signal amplitude, the digital inverters are able to preserve the phase information and provide very robust signal restoration up to their maximal operation frequency. Fig. 5 shows the simulation of the time-domain signal at fourteen 300 µm-spaced tap points on a 30 GHz LO bus. The two smallest-amplitude signals are the output of the first two buffers from the input of the bus: after these, signal levels are steady and adequately recovered at each stage, with no sensible degradation even after a large number of restorations. This LO distribution strategy appears to be very robust, and is limited only by the maximal operation frequency of ECL inverters available. This is, for large arrays, a frequency much higher than that reachable with linear IF buses: the combination of analog and digital design for IF and LO distributions allows faster RF output signals and larger array sizes than those achievable with analog-only design approaches.

3. Conclusion

While modern Si-based RF technologies allow integration of mm-wave transceivers, the integration of large arrays is difficult using IC architectures previously reported. We propose a mixed-signal row/column architecture, which bases its operation on the imposition of vertical and horizontal phase gradients. This greatly simplifies the array internal wiring, and reduces the system design to the design of adequate IF and LO buses. For the design of these two crucial subsystems, we propose a mixed signal approach. Given an adequate frequency plan, the robustness of digital design can be exploited for high-frequency non-linear LO signals, while linear analog design is necessary only for lower-frequency IF buses. The maximal output frequency and array size depends strongly on the application and available technology. We have presented preliminary simulations, based on libraries and technical data of an RF BiCMOS 0.18 µm commercial technology, which show that arrays larger than 12×12 are feasible in the $40 - 60$ GHz range.

References

1. R. Olsen, D. Rogers, and D. Hodge, "The aR^b relation in the calculation of rain attenuation," *Antennas and Propagation, IEEE Transactions on [legacy, pre - 1988]*, vol. 26, no. 2, pp. 318–329, 1978, 0096-1973.
2. A. Hajimiri, H. Hashemi, A. Natarajan, G. Xiang, and A. Komijani, "Integrated Phased Array Systems in Silicon," *Proceedings of the IEEE*, vol. 93, no. 9, pp. 1637–1655, 2005, 0018-9219.
3. A. Natarajan, A. Komijani, X. Guan, A. Babakhani, Y. Wang, and A. Hajimiri, "A 77GHz Phased-Array Transmitter with Local LO-Path Phase-Shifting in Silicon," in *IEEE International Solid-State Circuits Conference*, 2006, pp. 182–183.

International Journal of High Speed Electronics and Systems
Vol. 17, No. 1 (2007) 115–120
© World Scientific Publishing Company

TERAHERTZ EMISSION FROM ELECTRICALLY PUMPED SILICON GERMANIUM INTERSUBBAND DEVICES

N. Sustersic, S. Kim, P.-C. Lv[1], M. Coppinger, T. Troeger[2] and James Kolodzey

Department of Electrical and Computer Engineering, University of Delaware 140 Evans Hall, Newark, DE 19716, USA

In this paper, we report on current pumped THz emitting devices based on intersubband transitions in SiGe quantum wells. The spectral lines occurred in a range from 5 to 12 THz depending on the quantum well width, Ge concentration in the well, and device temperature. A time-averaged power of 15 nW was extracted from a 16 period SiGe/Si superlattice with quantum wells 22 Å thick, at a device temperature of 30 K and a drive current of 550 mA. A net quantum efficiency of approximately 3×10^{-4} was calculated from the power and drive current, 30 times higher than reported for comparable quantum cascades utilizing heavy-hole to heavy-hole transitions and, taking into account the number of quantum well periods, approximately four times larger than for electroluminescence reported previously from a device utilizing light-hole to heavy-hole transitions.

1. Introduction

Silicon Germanium (SiGe) optoelectronic devices are attractive because of compatibility with silicon integration and fabrication techniques. Silicon based optoelectronic devices for use in the terahertz (THz) region of the electromagnetic spectrum exhibit lower free-carrier and reststrahlen-band absorption than in III-V compound semiconductors. Because the nature of the bandgap is irrelevant for quantum well devices utilizing intersubband transitions, these transitions can be applied to fabrication of THz emitting and detecting SiGe quantum well devices. THz quantum well emitters based on intersubband transitions have the advantage of good coupling between the neighboring stages of the device. The quantum efficiency and output power scale with the number of active stages employed.

Quantum wells in the SiGe/Si material system may be realized by sandwiching thin SiGe layers between Si barriers. Due to the complex nature of the valence band, the SiGe quantum wells may confine light-holes, heavy holes, and spin-orbit split-off holes. The amount of overlap of the quantum well potentials for these different carriers depends on the Ge content. Transitions are possible between quantum confined states within just one band, i.e., purely between states of either the valence or the conduction band. The exact nature of the bandgap is irrelevant for a device utilizing such transitions, and this approach may therefore be applied to light emitter and detector devices fabricated using SiGe quantum well layers.

[1] P.-C. Lv is now with Intelligent Automation, Inc., 15400 Calhoun Dr., Suite 400, Rockville, MD 20855
[2] T. Troeger is now with Intel Corporation, 2200 Mission College Blvd., Santa Clara, CA 95052

2. Device Fabrication

For the THz emitters described here, fully strained commensurate SiGe layers were grown on Si (001) substrates by Molecular Beam Epitaxy (MBE). Using software that computes wavefunctions and energy levels based on a one-dimensional finite-element approach to the 8-band k·p approximation, the wavefunctions and energy levels were calculated.[3]

Three different samples with 16-period SiGe/Si superlattices were grown by solid source MBE at a substrate temperature of 400°C to minimize the effects of Ge segregation.[4,5] Sample 1 (SGC 396) had 16 Å thick wells and 30 Å barriers and was grown on a phosphorus-doped Si substrate with resistivity of 10 kΩ-cm. Samples 2 (SGC 420) and 3 (SGC 437) were grown on 1-10 Ω-cm p-type Czochralski (CZ) boron-doped Si substrates with sample 2 having 13.5 Å thick wells and 25 Å barriers, and sample 3 having 22 Å thick wells and 34 Å barriers. All samples grown had a germanium concentration in the wells of 30% with pure Si barriers. With the exception of a Si cap, which in sample 2 and 3 was doped with boron to a concentration of 5 x 10^18

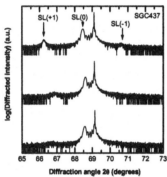

Fig. 1. X-ray diffraction data obtained from sample 3 (SGC 437). The angular region near the Si substrate peak showed superlattice satellite peaks and well-resolved Pendellösung fringes originating from dynamic X-ray effects such as multiple reflections at material boundaries.

Fig. 2. Cross-sectional TEM image taken from SGC 420. The 16 periods of 1.35 nm $Si_{0.70}Ge_{0.30}$ quantum wells and 2.5 nm Si barriers are clearly resolved.

cm^{-3}, the superlattice structures were not intentionally doped. Fig. 1 shows the X-ray diffraction data for sample 3 that was analyzed by software to yield crystallinity, strain, layer thicknesses, and composition. Fig. 2. shows a cross-sectional TEM image taken from sample 2.

For sample 2, device mesas of size 100 μm by 1850 μm and approximately 17 μm deep were fabricated using a multi-step reactive ion etching (MSRIE) process.[6] Metal contacts consisting of 200 Å Ti, 200 Å Pt, and 5000 Å Au were deposited using e-beam evaporation and a standard lift-off technique. For sample 1, a metal finger pattern 50-150 μm wide consisting of 200 Å Ti/2500 Å Au was deposited directly on the sample surface using a lift-off technique without etching. The fingers were spaced 70-350 μm and connected to a large common bond pad. The device area for this sample was several mm². For sample 3, and additional devices fabricated from sample 2, mesas approximately 200 × 200 μm in size and 2-3 μm deep were etched using RIE, utilizing a photoresist mask. The contacts were approximately 120 × 60 μm in size, leaving some of the top device mesa surface uncovered by metal in order to be able to investigate both top and edge emission.

For all devices, a large-area backside contact was achieved by evaporating Ti/Pt/Au or Ti/Au with thickness profiles the same as the top contacts. For electroluminescence measurements, all devices were mounted onto copper sample holders using thin indium sheets or conductive epoxy. For measurements at cryogenic temperatures, the sample holders were attached to the cold finger of an open-loop liquid helium cryostat. The topside metal pads were contacted by wedge bonding Au wires.

3. Electroluminescence Measurements

Electroluminescence spectra were measured by Fourier Transform Infrared Spectroscopy (FTIR) using a ThermoNicolet Nexus 870 spectrometer in the step-scan mode. The detector was a liquid helium cooled silicon bolometer, which is capable of detecting emission wavelengths from 16 μm to 500 μm. Duty cycles were chosen to prevent sample heating during the measurement, which would produce undesired blackbody radiation. At a pulse repetition rate of 413 Hz, trains of a varying number of pulses with widths ranging from 500 to 800 ns, and amplitude in the tens of volts were applied to obtain emission with minimal blackbody heating. A lock-in amplifier was tuned to the pulse train repetition frequency in order to collect radiation emitted from the device under test and reject ambient light. A recessed-cone blackbody radiator was used to calibrate the emitted power. The temperature dependence of the emission at 20 K and 30 K was studied using peak current pulses of 400 mA. The current versus voltage was measured at the same pulse conditions as the emission using an inductive probe and an oscilloscope.

4. Results and Discussion

The spectral features observed from sample 3 in the edge and top emission geometries consisted of boron impurity emission lines at low temperature. In top emission, at higher temperatures (>10 K), additional features were observed at shorter wavelengths than the impurity emission, which could not be replicated using a device without quantum wells. Measurements using devices with complete top metal coverage only revealed the boron impurity emission, regardless of the temperature. The features attributed to quantum well emission from sample 3 were only clearly resolved in top emission geometry.

Fig. 3 shows electroluminescence observed from sample 3 in top-emission geometry. By using a calibration carried out using the known spectral emission from a blackbody radiator, a time-averaged power of 15 nW was extracted for sample 3, at a device temperature of 30 K and a drive current of 550 mA in top-emission geometry, comparable in intensity with the impurity emission at that temperature and current.

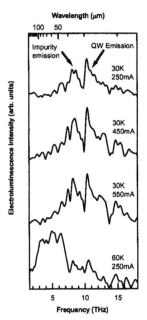

Fig. 3.   Electroluminescence observed from sample 3 (SGC 437) in top-emission geometry.   A peak attributed to impurity emission is seen centered around 8 THz along with a peak attributed to emission from the QW intersubband transition located between 11-12 THz.   At raised temperature (60-80 K), the peak shifts to approximately 5 THz.

A net quantum efficiency of approximately 3 x 10^{-4} was calculated from the power and the drive current, 30 times higher than reported for comparable quantum cascades utilizing heavy-hole to heavy-hole transitions[1] and, taking into account the number of quantum period repeats, approximately four times larger than for reported electroluminescence from a device utilizing light-hole to heavy-hole transitions.[2]

Features were observed from sample 3 in the range centered around 5 THz (~22 meV energy, 60 μm wavelength) in both top and edge emission when the sample temperature was raised to 50-60 K as in Fig. 3.   The peak intensity decreased when the temperature was increased further, but the line persisted to a temperature of 80 K.   Currently, it is not clear what causes this feature; the sharply defined spectral nature suggests that it is not due to blackbody radiation.   Because of the elevated temperature, the emission is unlikely to be due to boron intracenter transitions involving the Γ7 ground state.   It could be speculated that the transition causing this spectral feature to appear involves quantum well confined spin-orbit states, which had been disregarded for the calculations.

All samples had quantum well thicknesses and Ge contents that were intended to allow just two confined states in the wells, a light-hole and a heavy-hole state.   Such devices are expected to have advantages over commonly used designs where radiative transitions occur between two confined heavy-hole states.   As can be seen in Fig. 4, to access the longer wavelength portion of the THz gap, very large quantum well widths are necessary if one were to use two heavy-hole states, because the energy separation for such states is large for small quantum well thicknesses, and asymptotically approaches zero as the well width goes to infinity.

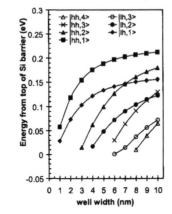

Fig. 4. Confined state energies for SiGe quantum wells with Ge content of 30% as a function of well width. SiGe QW THz emission is observed from the LH1 → HH1 radiative transition. Calculated using a one-dimensional finite-element approach to the 8-band k·p approximation

Conversely, the energy separation between the first confined heavy-hole and light-hole states increases from zero to a value determined by the difference of the heavy-hole and light-hole band offset with increasing quantum well width. For transitions between confined heavy hole states alone, light emission is only possible in edge emission geometry, because the momentum matrix element selection rules dictate that the light be polarized in the quantization (growth) direction. Transitions between light-hole and heavy-hole states result in light polarized in both the plane of the quantum wells and perpendicular to it. There is a light hole state located energetically between the two heavy hole states for all but the largest quantum well widths. Such a light hole state generates a parallel pathway for non-radiative phonon-assisted transitions, which lowers the overall quantum efficiency. The phonon-assisted transitions are expected to become more favorable for smaller energy separations between the two heavy-hole and the light-hole state. The momentum matrix element selection rules dictate that radiative transitions between two confined heavy-hole states may take place only at vanishing in-plane wave vector, whereas the subbands are populated to the Fermi wave vector. This selection rule may be relaxed somewhat under applied electric field, but in essence only allows a small fraction of the available carriers to undergo radiative transitions. For a light-hole to heavy-hole transition, the momentum matrix elements dictate that the in-plane wave vector needs to be non-zero, allowing radiative transitions for all carriers populating the initial subband.

Fig. 5. shows a summary of the observed THz electroluminescence data from the SiGe/Si quantum well devices with the emission frequency predicted by the 8-band k·p approximation for the LH1→HH1 radiative transition. The observed THz electroluminescence from the SiGe/Si quantum well devices shows excellent agreement with the calculated emission frequency based on quantum well width.

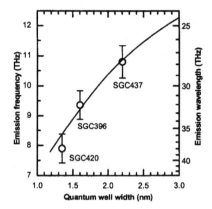

Fig. 5. Comparison of the observed peak center frequencies and wavelengths of the devices measured (open circles) with the emission frequency predicted for the LH1→HH1 radiative transition (solid line).

5. Conclusions

In summary, electroluminescence has been demonstrated at THz frequencies from devices utilizing SiGe/Si heterostructures consisting of thin SiGe layers sandwiched between Si barriers with transitions occurring between light-hole and heavy-hole states. A time-averaged power of 15 nW was achieved with a net quantum efficiency of 3×10^{-4}. The present results suggest that a terahertz laser based on SiGe/Si superlattice samples, optimized and possibly utilizing a greater number of active stages, should be attainable.

6. Acknowledgments

The authors would like to thank T. Adam, G. Xuan, and S. Ray. This work was supported by AFOSR program F49620-03-1-0380 and NSF Award DMR-0601920.

7. References

1. G. Dehlinger, L. Diehl, U. Gennser, H. Sigg, J. Faist, K. Ensslin, D. Grutzmacher, and E. Müller, *Science* **290**, 2277 (2000)
2. S. A. Lynch, R. Bates, D. J. Paul, D. J. Norris, A. G. Cullis, Z. Ikonic, R. W. Kelsall, P. Harrison, D. D. Arnone, and C. R. Pidgeon, *Appl. Phys. Lett.* **81**, 1543 (2002)
3. Quantum Semiconductor Algorithms, Inc., 5 Hawthorne Circle, Northborough, MA 01532-2711, U.S.A.
4. I. Bormann, K. Brunner, S. Hackenbuchner, G. Zandler, G. Abstreiter, S. Schmult, and W. Wegscheider, *Appl. Phys. Lett.* **80**, 2260 (2002)
5. D. J. Godbey, J. V. Lill, J. Deppe, and K. D. Hobart, *Appl. Phys. Lett.* **65**, 711 (1994)
6. T. N. Adam, S. Shi, S. K. Ray, R. T. Troeger, D. Prather, and J. Kolodzey, *Proc. IEEE Lester Eastman Conference on High Performance Devices*, 402 (2002)
7. P.-C. Lv, R. T. Troeger, T. N. Adam, S. Kim, J. Kolodzey, I. N. Yassievich, M. A. Odnoblyudov, and M. S. Kagan, *Appl. Phys. Lett.*, in press
8. H. Sakaki, "Fabrication of atomically controlled nanostructures and their device applications," In: Nanotechnology, G. Timp, ed., pp. 207-256, AIP Press, 1998

International Journal of High Speed Electronics and Systems
Vol. 17, No. 1 (2007) 121–126
© World Scientific Publishing Company

TERAHERTZ SENSING OF MATERIALS

G. Xuan, S. Ghosh, S. Kim, P-C. Lv, T. Buma, B. Weng, K. Barner and J. Kolodzey

Electrical and Computer Engineering Department, University of Delaware
Newark, DE 19716, USA
archyx@udel.edu

Biomolecules such as DNA and proteins exhibit a wealth of modes in the Terahertz (THz) range from the rotational, vibrational and stretching modes of biomolecules. Many materials such as drywall that are opaque to human eyes are transparent to THz. Therefore, it can be used as a powerful tool for biomolecular sensing, biomedical analysis and through-the-wall imaging. Experiments were carried out to study the absorption of various materials including DNA and see-through imaging of drywall using FTIR spectrometer and Time Domain Spectroscopy (TDS) system.

Keywords: Terahertz; DNA; drywall

1. Introduction

THz (~10^{12} Hz) lies between the Far-IR and the microwave frequencies, roughly from 300GHz to 10 THz, or about 1 to 40 meV in photon energy. Some materials interact strongly at THz frequencies, making them a great way to gain information that is otherwise not possible with X-Ray, optical or NMR [1]. Terahertz can also serve as a carrier wave in telecommunication applications for larger bandwidth [2]. These promising potentials attract intense R&D efforts worldwide.

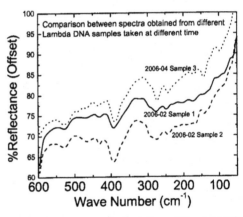

Fig 1. Three spectra obtained from different samples prepared using Lambda DNA. The spectra share most of the features even though they were taken at different times from different samples. It shows the reproducibility of the measurements is very good.

2. Experimental

In this paper, two experiments were conducted to investigate the potential of Terahertz in the DNA sensing and transport mechanism in random media.

2.1. *DNA sensing*

Label-free DNA detection has attracted tremendous research effort in recent years for its non-invasive, fast and low cost advantages [3-5]. In this study we explore the possibility of gaining information about the DNA through its interaction with Terahertz radiation. We use a Thermo Nicolet 870 FTIR spectrometer equipped with a liquid Helium cooled silicon bolometer (IRLabs Inc.) to study the absorption characteristics of DNA molecules. The 20bp and Lambda DNA were purchased from Invitrogen Inc. in a lyophilized power form. The 470bp and 1200bp DNA were synthesized at the University of Delaware. The DNA powder was then dissolved in DI water and 20μL of the solution was deposited with a precision micropipette on Au-plated Si substrates. The sample was subsequently vacuum-dried. The resulting DNA film has a thickness of about 10μm. We used a piece of Au-plated Si with no DNA film to serve as the reference background. The reflection spectrum was obtained from the ratio between the DNA film spectrum and the reference one. Before each measurement, the FTIR chamber was purged for two hours with Nitrogen to get rid of moisture. IR spectroscopy was carried out in the 600 – 50 cm^{-1} range (20 – 1.6 THz) at a resolution of 4 cm^{-1}.

The DNA spectra obtained from these films were reproducible and showed features that were unique to different DNA molecules (length and sequence). Figure 1 shows the reproducibility of the measurement. Using the same lambda DNA, three different samples were prepared and measured at different times. These spectra are nearly identical except that some resonance shows up stronger on one sample than on the other two.

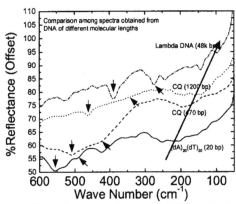

Fig. 2. Comparison among four types of DNA molecules of different lengths ranging from 20bp to 48kbp. These spectra have been vertically offset for clarity. Three resonant-absorption features exhibit length-related changes. These absorption peaks are marked by pointers in the graph.

Spectra were also obtained from DNA with varying lengths. Figure 2 shows the spectra for DNA molecules of 4 different lengths: 20 base pairs (bp), 470 bp, 1200 bp and 48 kbp. Woolard et. al. [5] published a spectrum on Herring sperm Tyhpe XIV DNA between 500 and 100cm^{-1} and showed similar spectra. In Fig. 2, from the bottom spectra to the top one, the DNA molecules became longer (more basepairs) and certain resonant-absorption peaks showed trend changes with longer molecule. For example, the 555cm^{-1} absorption peak in the 20bp spectra moves to 507cm^{-1} for 470bp, 463cm^{-1} for 1200bp and 390cm^{-1} for 48kbp. The 494cm^{-1} dip in 20bp spectra moves to the 363cm^{-1} for 1200bp and 274cm^{-1} dip for the 48kbp. This feature in 470bp spectra is greatly suppressed. The broad dip at 231cm^{-1} in the 20bp spectra moves to 190cm^{-1} for 470bp, 165cm^{-1} for 1200bp and 121cm^{-1} for 48kbp. To visualize this finding, Fig. 3 plots (log-log plot) the position of feature vs. DNA molecular length. Each curve shows each series of absorption peaks in different spectra that were believed to be length-related. In Fig. 3, it seems that all the features belonging to the 470bp DNA molecule show abnormal in the curve while the features from other three lengths DNA showed linear behaviors. In a log-log plot, linear behavior means one variable is proportional to the power of the other variable. The abnormal behavior from the 470bp spectrum could result from a number of possible reasons such as errors in determining the molecular lengths during electrophoresis. Curve-fitting results (less the non-linear points) in the equation in the form of

$$Peak = C_1 \cdot Length^{C_2}$$

and were shown next to each curve in Fig. 3. The thin lines in the graph were the result of the curve fits. The slopes of these three lines (C_2) were approximately in the same range (-0.0453, -0.0828 and -0.0757). Both constants (C_1 and C_2)

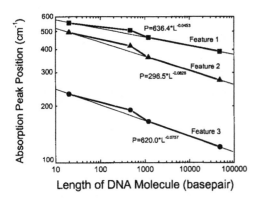

Figure 3. Log-log plot of position of absorption peaks versus lengths of DNA molecules. The thick line connects all the data points while the thin line is a perfect linear straight line connecting the resonant peaks from 20bp, 1200bp and 48kb.

are believed to be related to fundamental DNA molecular parameters such as length and sequence. If we could collect spectra from large quantities of different lengths DNA molecules combining the use of clustering methods, a correlation could be made between spectral features and DNA molecular information such as lengths and even hybridization states and sequence. No theories have been formulated to take into account of the DNA molecular length effect on the far-IR spectra yet.

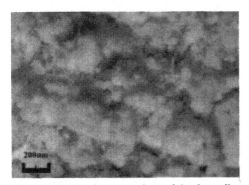

Figure 4. Optic microscope photo of the dry wall powder. It showed that the drywall particle size was around 200um.

2.2. *Drywall Transport of Terahertz*

Terahertz wave exhibits excellent properties that we could exploit for imaging purposes. Drywall is the most prevalent architectural material. The study of THz transport characteristics in drywall would provide useful information for THz imaging real-world applications. In addition, drywall is made up of many small particles and it is categorized as what is called "random media". Electromagnetic wave transport in random media receives much attention in the research community [6, 7] since they constitute an important part of our world: the rain, the fog, the clouds, the sandstorms, the building walls etc. are all random media. Thus the study of the THz photon transport mechanism in the drywall would also provide useful information for the basic understanding of the general electromagnetic wave transport mechanisms in random media.

Drywall has low absorption in the Terahertz range and its grain sizes are comparable to the THz wavelength. Figure 4 showed the microscopy image of the dry wall powder. It showed that the drywall particle size was on the order of 200 μm. The drywall material is primarily made from calcium sulphate plaster in its semi-hydrous form ($CaSO_4 \cdot \frac{1}{2}H_2O$). The plaster is usually mixed with various additives to increase mildew and fire resistance such as fiberglass and foaming agent. A series of drywall samples of different thicknesses were prepared. The thick paper skin covering both sides of the drywall has been carefully removed.

For the broad band THz source we used a Time-domain Spectroscopy (TDS) system, which spans from 0.2 to 1 THz (1.5mm – 300um).

We placed each drywall sample normal to the THz beam and record the transmittance. The time-domain transmittance data from each sample was Fourier-transformed to obtain the transmission spectrum data in the frequency domain.

Scattering and absorption are the two major mechanisms affecting photon transport in materials. The drywall material shows relatively low absorption in the frequency range we used (0.2 – 1 THz). If reflection and scattering are negligibly small, the absorption coefficient is given by $\alpha = -\ln(T)/L$ (T – transmittance, L – sample thickness). Drywall has minimal reflection in the THz region. And let's assume in the extreme case, where absorption was the only cause for signal attenuation through dry wall samples, then at 0.25 THz, $\alpha = 1.2 cm^{-1}$, at 0.4THz, $\alpha = 2.8 cm^{-1}$; and at 0.6THz, $\alpha = 5 cm^{-1}$. Materials with absorption coefficient of only a few cm^{-1} are considered weak absorption material. The weak absorption of dry wall makes it easier to observe the effect of scattering on the photon transport properties.

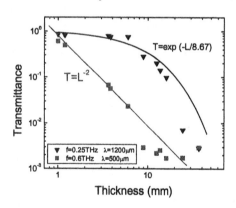

Figure 5 Transmittance vs. drywall sample thicknesses for two different frequency THz wave: 0.25 and 0.6THz. The transmittance data was fitted with an exponential function for the 0.25THz and the quadratic function for the 0.6THz.

Figure 5 shows the transmission vs. sample thickness at 0.25 and 0.6 THz in the log-log plot. At 0.25THz, the wavelength (1200μm) is much more than the drywall particle size (~200μm) and thus the scattering plays little role while the absorption dominates. At 0.6 THz, the wavelength (500μm) is approaching the drywall particle size (~200μm) and thus (from Mie-scattering) we expect to see the scattering starts to dominate. As seen from Figure 5, at 0.25THz the attenuation is mainly due to absorption and thus the data appeared exponential: $T=\exp(-L/8.67)$. At 0.6THz, with more scattering affecting the photon transport, the transmittance is drastically reduced. And the 0.6THz transmittance data appeared to be quadratic in relation to the inverse thickness: $T=L^{-2}$. This quadratic relationship indicates the onset of photon localization. [8] If we could obtain larger size particles in the drywall material, at 0.6THz the photon would probably be localized already, in which case the material would be opaque to the THz wave. This result showed that for THz through-the-wall sensing applications, lower THz frequency wave are strongly preferred for its higher transmission.

3. Summary

Measurements on DNA of varying length has been reliably performed using FTIR in the 600-50 cm^{-1} spectral region and several features have been identified as to be DNA molecular length related. Photon transport in random medium was explored using THz through drywall. We have demonstrated that the higher frequency THz waves has exhibited the onset of photon localization inside drywall material, and that the lower frequence THz transmits better through drywall.

4. Acknowledgements

This work was supported by AFOSR program: F49620-03-1-0380, NSF Award, DMR-0601920.

References

1. P. H. Siegel, *Terahertz Technology*, IEEE Transactions on Microwave Theory and Techniques, **50**(3) (2002) 910
2. D. Woolard, *Terahertz electronic research for defense: Novel technology and science*, in 11th Int. Space Terahertz Tech. Symp., Ann Arbor, MI, May 1-3, 2000, pp. 22-38
3. E. Souteyrand, J. P. Cloarec, J. R. Martin, C. Wilson, I. Lawrence, S. Mikkelsen, and M. F. Lawrence, *Direct Detection of the Hybridization of Synthetic Home Oligomer DNA Sequences by Field Effect*, J. Phys. Chem. B **101** (1997) 2980
4. G. Xuan, J. Kolodzey, V. Kapoor and G. Gonye, *Characteristics of field-effect devices with gate oxide modification by DNA*, Appl. Phys. Lett. **87** (2005) 103903
5. D. L. Woolard, T. R. Globus, B. L. Gelmont, M. Bykhovskaia, A. C. Samuels, D. Cookmeyer, J. L. Hesler, T.W. Crowe, J. O. Jensen, J. L. Jensen, and W. R. Loerop, *Submillimeter-wave phonon modes in DNA macromolecules*, Phys. Rev. E, **65** (2002) 051903

6. S. Kumar and C.L. Tien, *Dependent Absorption and Extinction of Radiation by Small Particles*, J. Heat Transfer-Transactions of the ASME, **112** (1990) 178

7. M.I. Mishchenko, L.D. Travis, and A.A. Lacis, *Scattering, Absorption, and Emission of Light by Small Particles*, Cambridge University Press, New York, 2002

8. D.S. Wiersma, P.Bartolini, A. Lagendijk, and R.Righint, *Localization of Light in a Disordered Medium*, Nature, **390** (1997) 671

Section III.
Silicon and SiGe Devices

International Journal of High Speed Electronics and Systems
Vol. 17, No. 1 (2007) 129–141
© World Scientific Publishing Company

NEGATIVE BIAS TEMPERATURE INSTABILITY IN TIN/HF-SILICATE BASED GATE STACKS

N. A. CHOWDHURY, D. MISRA AND N. RAHIM

Electrical & Computer Engineering Department, New Jersey Institute of Technology, University Heights, Newark, NJ 07102, USA, nac5@njit.edu

This work studies the effects of negative bias temperature instability (NBTI) on p-channel MOSFETS with TiN/HfSi$_x$O$_y$ (20% SiO$_2$) based high-κ gate stacks under different gate bias and elevated temperature conditions. For low bias conditions, threshold voltage shift (ΔV_T) is most probably due to the mixed degradation within the bulk high-κ. For moderately high bias conditions, H-species dissociation in the presence of holes and subsequent diffusion may be initially responsible for interface state and positively charged bulk trap generation. Initial time, temperature and oxide electric field dependence of ΔV_T in our devices shows an excellent match with that of SiO$_2$ based devices, which is explained by reaction-diffusion (R-D) model of NBTI. Under high bias condition at elevated temperatures, due to higher Si-H bond-annealing/bond-breaking ratio, the experimentally observed absence of the impact ionization induced hot holes at the interfacial layer (IL)/Si interface probably limits the interface state generation and ΔV_T as they quickly reach saturation.

Keywords: NBTI; Hafnium Silicate.

1. Introduction

Down scaling of MOS field effect transistor (MOSFET) sizes, specifically oxide thickness below 1.6nm increases transistor leakage current to levels unacceptable for the low power applications. An attractive solution is to replace SiO$_2$ with high-κ dielectric materials while retaining the standard MOSFET design. However, it is known that when advanced integrated circuits operate at elevated temperatures negative bias temperature instability (NBTI) becomes a serious degradation mechanism.[1] Hence, it is being thoroughly studied specifically for low dimension devices. In p-channel MOSFETS NBTI occurs when the device is operating at a higher temperature and is under inversion condition i.e. gate is negatively biased with respect to substrate. At this condition defects are generated resulting in threshold voltage instability. For SiO$_2$ case, interface state generation, triggered by Si-H bond breaking at Si/SiO$_2$ interface is the point of major concern. One of the widely-used realistic NBTI models is based on reaction-diffusion (R-D) theory, which basically focuses on time dependent net increase in the number of interface states, N_{it} as the competing processes of bond-breaking and bond-annealing take devices, it could not fully explain the observed NBTI results in Hf-based high-κ gate stacks.[2–5]

Observed threshold voltage shifts in high-κ devices suggest that bulk trap generation exceeds that of interface states generation in most of the studies,[2-5] specifically in cases of low stress biases. Aoulaiche *et al* reported that both fast and slow states were generated and subsequently recovered by applying low bias (≤ –2 V) at elevated temperature on TaN/HfSiON gate stacks.[2] However, when –1.5 V of negative bias temperature (NBT) stress at 125°C was applied on poly-Si gate/HfO₂ devices, such recovery was not observed by Houssa *et al* and it was attributed to possible generation of hydrogen-related centers within the bulk.[3] Fujieda *et al* simultaneously observed electron trapping, interface state generation and positive charge build-up when they applied NBT stress bias of –1.5 to –2.5 V on poly-Si gate/ HfSiON devices.[4] Harris *et al.* reported that interface state generation was negligible when –2 to –3 V of NBT stress was applied on TiN/HfSi$_x$O$_y$ devices and V_T instability was attributed on both shallow and deep electron traps within the bulk high-κ.[5] It is obvious that a uniform NBTI induced degradation scenario can not be extracted from these studies, most probably due to the diversity in high-κ gate stacks processing conditions. Therefore, NBTI studies need to be carried out on one-to-one basis on individual high-κ gate stacks.

In this paper, we reported that mixed degradation due to both electron and hole trapping within the bulk high-κ dominates as constant voltage stress (CVS) with low negative gate bias is applied on pMOSFETS with metal organic chemical vapor deposited (MOCVD) TiN/HfSi$_x$O$_y$ (20% SiO₂) based gate stacks with effective oxide thickness (EOT) of ~1.8 nm. As gate bias is increased to moderately high levels, initially both interface state and bulk trap generation becomes significant, which results in power-law dependence of negative ΔV_T. Comparison of time, temperature and oxide electric field dependence with R-D theory based phenomenological models for SiO₂ implies that dissociation and diffusion of H-species in the presence of holes at Si/IL interface are probably responsible for mostly positively charged trap generation induced ΔV_T. When the applied gate bias is high, ΔV_T, due to positive charge build-up, quickly saturates at elevated temperatures. Higher Si-H bond-annealing/bond-breaking ratio due to the absence of impact ionization induced hot holes at IL/Si interface may be responsible for the latter observations.

2. Experimental

Hafnium silicate (HfSi$_x$O$_y$ –20% SiO₂) film and TiN metal gate were deposited by MOCVD technique[6] on both n- and p-type Si substrates after ozone treatment had been performed for the pre-dielectric deposition cleaning, which resulted in ~10Å of chemical oxide growth at the dielectric and Si substrate interface.[6] Both n- and p-channel MOSFETs were fabricated using the standard CMOS process flow. Using HRTEM, the physical thickness has been measured to be 4.5nm including an interfacial layer (IL) of 1 nm.[6] These devices were further subjected to NH₃ PDA at 700°C for 60 secs to improve the leakage performance. The physical characterization details can be found elsewhere.[6] An EOT of 1.8-2 nm was estimated from the high-frequency (1 MHz) *C-V* measurements after the quantum mechanical corrections.[6] CVS and SHH stress were applied on pMOSFETs using HP4156B semiconductor parameter analyzer. Micromanipulator hot stages were used for measurements at elevated temperatures.

3. Results and Discussions

In this work, a large number of fresh pMOSFETs were used. V_T and sub-threshold swing, S were found to be within −0.78 to −0.8 V and 85 to 90 mV/dec range in the fresh

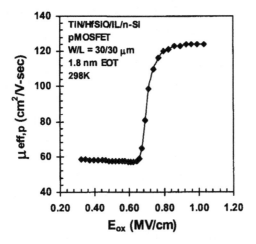

Fig.1. $\mu_{eff,p}$ vs. E_{ox} for a fresh pMOSFET at room temperature.

devices. Characteristics of the fresh devices were quite uniformly distributed. The effective hole mobility, $\mu_{eff,p}$ of a typical fresh device is shown in Fig.1 as a function of oxide electric field. It may be noted that for surface doping density of ∼1×10^{18} cm^{-3}, $\mu_{eff,p}$ in our devices is around 80% of SiO$_2$ based pMOSFETs.7

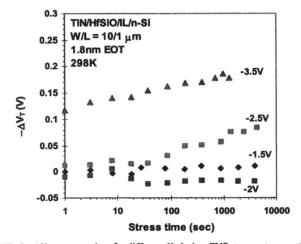

Fig.2. ΔV_T vs. stress time for different V_g during CVS at room temperature.

CVS with gate bias, V_g in the range of –1.5 to –3.5 V was applied on pMOSFETS with W=10 μm and L= 1 μm at room temperature (RT). Fig.2 shows that for high V_g conditions, positive charge trapping dominates. It is further observed that that ΔV_T finally saturates. For low bias conditions, mixed degradation occurs due to both positive and negative charge trapping within the bulk high-κ. This is consistent with our previous work with the same gate stacks,[8,9] where we reported the presence of both deep electron and hole traps within the bulk high-κ. Specifically, for –2 V, electron trapping dominates at RT. I_d-V_g plots in Fig.3 for –2 V shows almost no change in S. As $\Delta S \propto \Delta D_{it}$,[4] interface state generation is negligible at RT for –2 V.

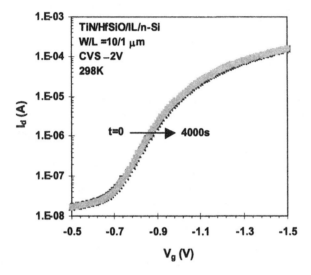

Fig. 3. *I_d-V_g plots for CVS with V_g = –2 V at RT.*

In order to understand the role of holes in interface state generation during gate injection in our devices, CVS was applied with and without non-zero substrate bias, V_b for V_g = –2.5 V and –3.5 V conditions at RT. For the sake of equivalence, V_b was kept numerically equal to V_g. Carrier separation technique, as shown in the figure 4(a), shows that impact ionization induced reversal of the polarity of the source/drain current, $I_{S/D}$[10,11] did not occur even under V_g / V_b = – 3.5 V/ +3.5 V stress condition. It is consistent with results from p[+]-poly gate/n-Si structures, where V_g = –3.5 V was found to be the threshold for the impact ionization.[10] As TiN is a mid-gap material, threshold in our case is $V_g \approx$ – 4 V.

Increment in I_{gate} can be attributed to the increased hole trapping at the pre-existing or stress-induced hole traps. Increase in trap-assisted tunneling (TAT) due to stress-induced traps may be another cause of increase in gate current. Increase in $I_{S/D}$ can be also attributed to increased TAT.

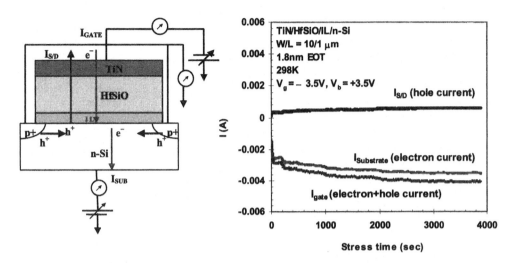

Fig.4.(a) Carrier separation technique as CVS is applied with negative gate and nonzero substrate bias.
(b) Current vs. stress time showing electron and hole injection during CVS with nonzero substrate bias.

Figs. 5(a) and (b) show $\Delta S/S_0$ and ΔV_T during CVS under both zero and non-zero bias conditions. It is obvious from Fig. 4(a) that enhanced presence of holes due to $V_b > 0$ increases interface state generation. It is further observed from log-log plots that $\Delta S/S_0$ follows t^n power-law dependence. The value of exponent, $n \approx 0.2$ for $V_g = -2.5$ V under both $V_b = 0$ V and $V_b = 2.5$ V conditions in our case. This value of n is related to Si-H bond breaking in the presence of low energy holes.[12] Dominance of hot holes results in, $0.2 < n < 0.5$.[12] This is consistent with our earlier conclusion that impact ionization induced hot holes were not generated during gate injection. For $V_g = -3.5$ V, initial increase in interface state generation was significant. However, the absence of hot holes retarded Si-H bond breaking rate under $V_b = 0$ V condition. Therefore, a large increase in interface state generation required to sustain $n \approx 0.2$ was not possible. Hence, it initially increased with $n \approx 0.1$ and finally tended to saturate. Nevertheless, $n \approx 0.2$ could be retained due to the presence of increased number of low energy holes under $V_b = 3.5$ V condition. It may be noted that these observations are also supported by our previous work with MOS capacitors,[13] where it was shown that for $V_g = -4$ V under CVS and $V_b = 4$ V under SHH stress, ΔV_{FB} increased with $n \approx 0.5$, which is characteristic of hot hole generation.

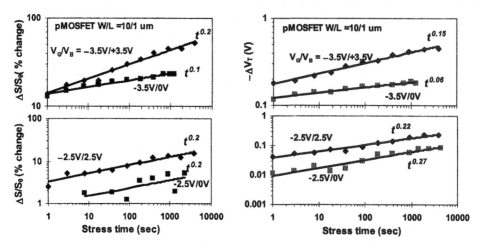

Fig. 5. (a) $\Delta S/S_0$ and (b) ΔV_T for both zero and non-zero substrate bias conditions during CVS at RT.

It can be observed from Fig. 5(b) that $|\Delta V_T|$ also follows t^n power-law dependence under all stress conditions. For $V_g = -2.5$ V stress level, n for ΔV_T is greater than that found for $\Delta S/S_0$ under both zero and non-zero V_b conditions. This indicates that besides interface states, positively charged bulk trap generation also take places.[2] For $V_g = -3.5$ V stress level, however, n for $|\Delta V_T|$ is lower than that for $\Delta S/S_0$ under both V_b conditions. Besides interface states and positively charged bulk trap generation, electron trapping may be slowing the increase of $|\Delta V_T|$, so that finally it saturates. This is consistent with our earlier studies,[9] where we observed electron trapping at deep electron traps is significantly high for high negative bias stress levels.

For stress at elevated temperature of 398K (125°C), it can be observed from Fig. 6 that ΔV_T shows saturation for high bias conditions. However, for low bias conditions mixed degradation is still obvious.

To understand temperature dependence on degradation further, stress was carried out for both low and high stress conditions for extended period of time at different Elevated temperatures. It is obvious from Fig. 7(a) that positive charge trapping is thermally activated for −2 V. Initially positive charge trapping increases with time for each temperature condition; however it reaches saturation after ~300 secs of stress. Similarity in the patterns in Figs. 7(a) and (b) suggests that interface state generation and ΔV_T are highly correlated at elevated temperature conditions.

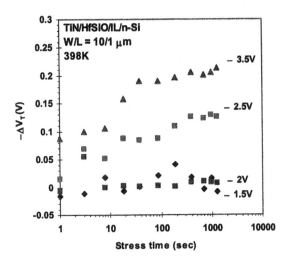

Fig.6. ΔV_T vs. stress time for different V_g at 398K (125°C) applied on pMOSFETS.

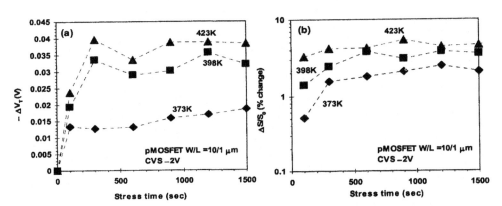

Fig. 7. (a) ΔV_T and (b) $\Delta S/S_0$ for CVS with −2 V of stress level at elevated temperatures.

For −3.5 V, positive charge trapping reaches saturation earlier at elevated temperatures of 373K (100°C) and 398K (125°C), but initially it shows temperature dependence as shown in Fig. 8(a). $\Delta S/S_0$ also shows a similar behavior (Fig. 8(b)), which further implies the effect of the interface state generation on ΔV_T.

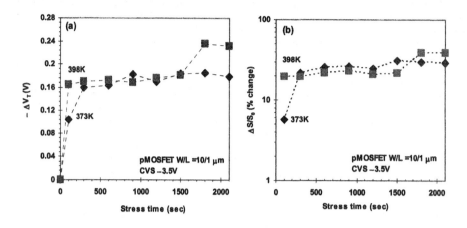

Fig. 8. (a) ΔV_T and **(b)** $\Delta S/S_0$ for CVS with –3.5 V of stress level at elevated temperatures.

It is obvious from Figs. 7 and 8 that at the initial stage of the stress, ΔV_T and $\Delta S/S_0$ depends not only stress level but also on temperature. In order to further understand these effects, ΔV_T and $\Delta S/S_0$ are plotted as a function of electric field (E_{ox}) for different elevated temperature conditions in Figs. 8(a) and 8(b). Here, $E_{ox} = (V_g - V_{FB} - \Psi_S)/EOT$, where Ψ_S

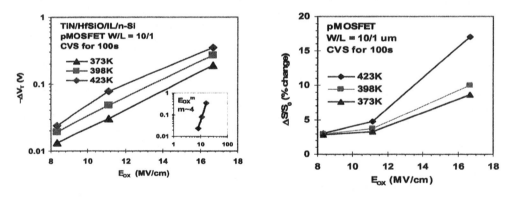

Fig. 9. (a) ΔV_T and **(b)** $\Delta S/S_0$ vs. E_{ox} after initial 100s of CVS under different elevated temperatures conditions. **(Inset)** ΔV_T vs. E_{ox} at 423K (150°C).

is the surface potential. ΔV_T and $\Delta S/S_0$ were measured after 100s of uninterrupted CVS. For each field and temperature condition, a fresh device was used (total: 9). It can be observed that ΔV_T strongly depends on field and temperature conditions, specifically during the initial period of stress. Inset of Fig. 9(a) shows that ΔV_T shows E_{ox}^{m} power-law dependence and for 423K $m \sim 4$. For other temperatures, m is in the same range.

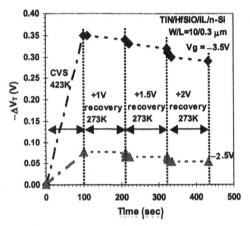

Fig.10. ΔV_T vs. time during CVS at 423K (150°C) with $V_g = -3.5$ V and -2.5 V, and post-stress recovery under different positive gate bias conditions.

To understand post-stress recovery, injection of substrate electrons was done at different positive biases as shown in Fig. 10. But it fails to neutralize $|\Delta V_T|$ for both -2.5 V and -3.5 V stress conditions, which suggests that H-species may be responsible for positively charged trap generation within the bulk.[3]

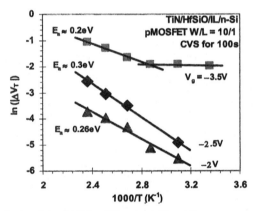

Fig. 11. Arrhenius plots of ΔV_T for different gate bias. ΔV_T was measured after initial 100s of CVS.

Arrhenius plots of ΔV_T for different negative gate bias is shown in Fig. 11. ΔV_T was measured after initial 100s of CVS with different stress level. For a particular stress and temperature condition, a fresh device was used to avoid the effects of the residual trapping. It is obvious that positive charge trapping is thermally activated in our gate stack, specifically at the initial stage of stress. For $V_g = -3.5$ V stress level, saturation of ΔV_T took place at less elevated temperatures. However, as temperature is raised thermal activation of positive charge build-up became obvious. Activation energies were found to be lying within 0.2 to 0.3 eV range.

Reaction-diffusion model of NBTI is based on net positive increment of interface states, N_{it} as two competing processes of Si-H bond breaking and annealing occur simultaneously.[14] The following equation describes this:[1]

$$\frac{dN_{it}}{dt} = k_f \left(N_0 - N_{it} \right) - k_r N_{it} N_H^{(0)} \quad \ldots\ldots\ldots\ldots(1)$$

Here, k_f / k_r: bond breaking/annealing rate, N_0: number of Si-H bond density prior to degradation, $N_H^{(0)}$: H-species density at Si/SiO$_2$ interface. Si-H bond breaking/annealing at the presence of holes is shown by the following electrochemical reaction:

$$Si - H + h^+ \Leftrightarrow Si \bullet + H^0 \ldots\ldots\ldots\ldots\ldots\ldots(2)$$

Recent studies[12] show that energy of hole plays a role in bond-breaking and post-stress bond-annealing. For low energy holes, it is mostly Si$_3$≡Si–H bonds are broken, and ΔN_{it} increases with a power-law exponent, $n \approx 0.2$ as stated earlier. Moreover, a fraction of the broken Si$_3$≡Si– bonds anneal after stress. For hot holes, besides Si$_3$≡Si– species, ≡Si–O– defects are generated, which does not anneal after stress and raise the value of n to 0.2 to 0.5.

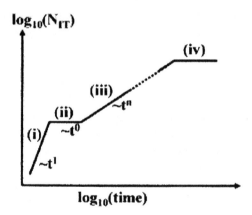

Fig. 12. $_{it}$ vs. stress time as predicted by R-D model (From Ref. 1).

Closer inspection of R-D model, as shown in Fig. 12,[1] reveals that initially (i) N_{it} increases by t^1 as $N_0 \gg N_{it}$. (ii) When bond-breaking = bond-annealing, N_{it} increases by t^0. (iii) When H-species diffusion into the oxide dominates, N_{it} increases by t^n. (iv) Finally, when $N_{it} = N_0$, interface state generation stops.

It is also possible to explain time, temperature and field dependence of ΔV_T in a compact form, specifically for region (iii) in Fig. 12. Phenomenological description of this dependence can be expressed in the following way:[14]

$$\Delta V_T = CE_{OX}^m \exp\left(\frac{-E_a}{k_B T}\right) t^\alpha \quad \ldots\ldots\ldots\ldots\ldots\ldots(3)$$

Here, C is a constant and k_B is Boltzmann's constant. For Si/SiO$_2$, it was found that $m \sim 3-4$, $E_a \sim 0.1-0.2$ eV and $\alpha \sim 0.2-0.25$. Comparison with our experimental results shows excellent match. Therefore, R-D model may be explored to interpret NBTI effects in our devices.

Initially our devices may not be in region (i) of Fig. 12 since our initial $D_{it} \sim 1 \times 10^{12}$ (cm^{-2}.eV^{-1}). Our experimental results in Fig. 5(a) show that initially our devices reside in region (iii). Moreover, it is supported by phenomenological model in Eq. (3). While in region (iii), ΔV_T increases initially due to Si–H bond breaking, which results in both interface states and diffused H–species induced bulk trap generation. Finally, however, ΔV_T tends to saturate. Migration to region (iv) due to the usurping of all available Si–H bonds may be considered. But, Fig. 8 shows that if stress-time is increased, more bond-breaking may occur, which increases both $\Delta S/S_0$ and ΔV_T for $V_g = -3.5$ V at 398K. Therefore, the breaking of all Si-H bonds may not be the cause of the saturation.

As stated earlier, the lack of hot holes during CVS results in higher post-stress anneal of broken bonds,[12] which may limit the increase of ΔN_{it}, that is, $\Delta S/S_0$ and ΔV_T. We have experimentally observed that the signatures of the impact ionization induced hot hole generation are absent even under the extreme stress conditions studied in this work. Moreover, the value of the power-law exponent, $n \approx 0.2$ for $\Delta S/S_0$ also implies that the low energy hole induced Si–H bond breaking dominates in our case. Based on this, we may argue that the bond-annealing due to the low energy hole induced Si–H bond breaking, the number of which declines with the progress of stress, is possibly the cause of the tendency of $\Delta S/S_0$ and ΔV_T to saturate under a particular bias temperature stress condition.

To further discuss our results we have applied CVS on ALD deposited TiN/HfO$_2$ based gate stacks (with 26A HfO$_2$ and 11A of IL) at room and elevated (398 K) temperatures. It is obvious from Fig. 13 that dependence of ΔV_T on stress time follows the power-law with exponent ~ 0.1. This low value of n may be due the increased negative charge trapping within the bulk oxide.[5] Moreover, ΔV_T depends on temperature and gate bias, i.e., electric field conditions. Furthermore, inset of Fig. 13 shows that positive charge trapping increases with non-zero substrate bias. These results conform to our earlier observations of the NBTI effects on TiN/Hf-silicate based gate stacks.

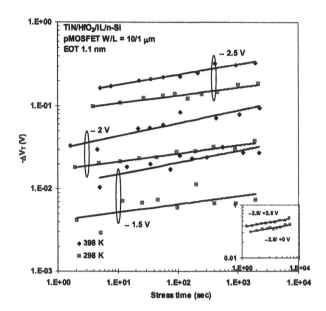

Fig. 13. ΔV_T vs. stress time for CVS applied on pMOSFETs with TiN/HfO$_2$ based gate stacks under different negative bias and temperature (RT and 398 K) conditions. **(Inset)** ΔV_T vs. stress time plots for CVS under zero and non-zero substrate bias conditions at RT.

4. Conclusions

NBTI effects under different bias and temperature conditions were studied for TiN/HfSi$_x$O$_y$ (20% SiO$_2$) based high-κ gate stacks. For low bias conditions, mixed degradation due to both electron and hole trapping within the bulk high-κ mostly dominates ΔV_T. Interface state generation, observed from change in sub-threshold slope, $\Delta S/S_0$, was found to be negligible. For moderately high to high stress levels, initially Si-H bond breaking induced interface states and diffused H-species induced bulk trap generation dominate. Initial temperature, time and oxide electric field dependence shows excellent match with that of R-D based NBTI model. Carrier separation technique shows that impact ionization induced hot hole generation, signature being the reversal of the polarity of source/drain current during stress, was not observed. This possibly results in a higher bond-annealing/bond-breaking ratio as, with the progress of the stress; a less number of bonds are available to be broken at the presence of the low energy holes. This may be responsible for the observed saturation of the interface state generation and ΔV_T under high bias temperature stress conditions.

5. Acknowledgments

We thank Dr. B. H. Lee and Dr. R. Choi of SEMATECH, Austin, Texas for research collaboration. This work was partially supported by a National Science Foundation grant (#ECS-0140584).

References

1. H. Kufluoglu and M. A. Alam, *IEEE TED 53*, 1120 (2006).
2. M. Aoulaiche, M. Houssa, R. Degraeve et al., *Microelec. Engnr. 80*, 134 (2005).
3. M. Houssa et al., *J. ECS 151*, 288 (2004).
4. S. Fujieda et al., *Jap. JAP 44*, 2385 (2005).
5. H. R. Harris et al., *43rd IRPS*, 2005, p.80.
6. P. S. Lysaght *et al.*, *Proc. of MRS*, 747, 2003, p. 133.
7. H. C. Casey, *Devices for Integrated Circuits* (John Wiley and Sons, New York, 1999), p.75.
8. D. Misra and N. A. Chowdhury, *Second International Symposium on Dielectrics for Nanosystems: Materials Science, Processing, Reliability, and Manufacturing*, 209th Electrochemical Society Meeting, Denver, Colorado, 2006.
9. N. A. Chowdhury, P. Srinivasan and D. Misra, *Physics and technology of High Dielectric Constant Gate Stacks- III, 208th Electrochemical Society Meeting*, Los Angeles, California, 2005.
10. Y. Shi, T. P. Ma, S. Prasad and S. Dhanda, *IEEE TED 45*, 2355 (1998).
11. S. Mahapatra, P. B. Kumar, and M. A. Alam, *IEEE TED 51*, 1371 (2004).
12. D. Varghese, S. Mahapatra, and M. A. Alam, *IEEE EDL 26*, 572 (2005).
13. N. A. Chowdhury, P. Srinivasan, D. Misra, R. Choi and B. H. Lee, *Proceedings of 2nd SEMATECH International Symposium on Advanced Gate Stacks Technology*, Austin, Texas, 2005, p.77.
14. M. Houssa, G. Pourtois, M. M. Heyns, and A. Stesmans, *J. of Physics: Cond. Matter 17*, S2075 (2005).

International Journal of High Speed Electronics and Systems
Vol. 17, No. 1 (2007) 143–146
© World Scientific Publishing Company

POWER ADAPTIVE CONTROL ON DENSE CONFIGURED SUPER-SELF-ALIGNED BACK-GATE PLANAR TRANSISTORS

HAO LIN, HAITAO LIU*, ARVIND KUMAR*, UYGAR AVCI, JAY S. VAN DELDEN*, SANDIP TIWARI*

*School of Applied & Engineering Physics, *School of Electrical & Computer Engineering, Cornell University,
Ithaca, New York 14850, USA
lh77@cornell.edu*

ARVIND KUMAR

*IBM T. J. Watson Research Center,
Yorktown Heights, New York, USA*

The works examines the power adaptive control of back-gated planar transistor employing a novel super-self-alignment design. The availability of such back-gate provides additional freedom for the improvement of the transistor performance of higher switching speed and lower power consumption. Both of these two capabilities are investigated through the studies of transistor's output and transfer characteristics.

Keywords: Super-Self-Aligned, Back-gate, Transistor, Power Adaptive Control.

1. Introduction

As silicon CMOS technology emerge in nanoscale regime, a robust frame work for technology that emphasizes power and configurability while achieving speed, noise margin, reproducibility and reliability is highly desired[1]. Power dissipation and complexity of dense device interconnectivity are major barriers to higher computing power. The idea of double gate transistor was proposed to provide excellent short channel electrostatic control for low power application[2-3]. With additional freedom gained from independent back-gate and front-gate access, higher circuit performance can be achieved even with back-gate misalignment[4]. Self-aligned planar back gate transistor is required to generate ultimate device and circuit performance. Device technology of such a structure is considerably more complicated. A reproducible approach to the fabrication of super-self-aligned back-gate/double-gate transistors with additional feature of buried interconnection have now been demonstrated[5]. This technique also makes dense circuit configurability for adaptive power control feasible, and will be described in this work.

2. Device Fabrication

A combination of CMOS front end processing innovations featuring wafer bonding, oxide isolation combined with thin silicon channel control, solid-state junction diffusion and chemical mechanical polishing (CMP) for wafer level planarization highlight the novelties of this approach[5]. First, such approach employs a process similar to

SMARTCUT technology to transfer interconnected gate patterns onto a handle wafer to form an oversized back-gate SOI (silicon-on-insulator) structure. A dense circuit configuration can benefit from it such that interconnect can be made both on top of the device and buried underneath it inside the SOI BOX (buried oxide) region. Second originality comes from a series insulators (silicon oxide and silicon nitride) caps and sidewall spacer formation to achieve super-self-alignment between back-gate and front gate. It allows the formation of thin channel as well as thick source/drain to minimize contact series resistance. These features are in accordance with the high speed circuit requirement to maximize switching speed and minimize circuit RC delay. Due to this super-self-alignment process, device source/drain region are made by heavily doped polysilicon deposition which can selectively form either p-type or n-type transistor with different dopants. Finally, strain can be introduced through back-gate oxidation to change channel carrier mobility in order to further enhance device performance.

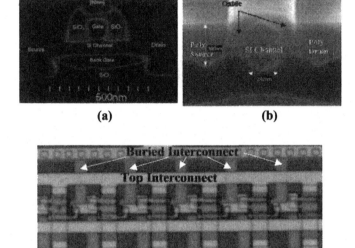

(a)   (b)

(c)

Fig. 1. Cross section SEM image of (a) back-gate transistor of 90nm nominal gate length and 70nm back-gate oxidation and (b) back-gate transistor of 300nm nominal gate length and 400nm back-gate oxidation; (c) Top and buried interconnects in circuit made of back-gated devices from top view optical micrograph.

Cross section images of the back-gate transistors are taken by scanning electron microscope in Fig. 1 (a-b). Devices of gate length as small as 90 nm have been fabricated. Different back-gate oxidation is applied to obtain different back-gate recess and introduce strain into silicon channel. Simple circuit made of series connected inverters in Fig. 1(c) show buried interconnects among devices in additional to interconnections on top. These buried interconnects can help alleviate the problem of ever increasing complication of circuit wiring for ULSI (ultra-large-scale-integration).

3. Device Characterizations

Device made on SOI wafer provide a low power solution to the escalating power consumption on silicon electronics. Especially the fully depleted ultra-thin body SOI has good electrostatic control of channel conduction to obtain better sub-threshold swing and higher on-off ratio of the transistor. This translates into improved short channel effect, which leads to faster switching speed and lower static power consumption. However, such device structure suffers from the loss of the threshold voltage control due to its undoped thin body and electric field coupling between channel and drain contact. In order to overcome these two difficulties, back-gated SOI structure is proposed[6]. The availability of this additional gate at the bottom of the channel reinforces channel electrostatic control to gain effective shielding of drain electric field and threshold voltage V_T modulation. Therefore device can be dynamically back-gate-biased at desired state to either minimize static leakage current or to achieve fast device switching.

N-channel and p-channel transistors, both employing n$^+$ gates have been fabricated and characterized to study the dynamic control of device through back-gate. Their output characteristics show either front or back channel conductions can be largely affected by the back-gate bias (Fig. 2 (a-b)). Large drive current as much as 800 µA/µm at V_{DS} =1 V has been seen on nFET and 100 µA/µm on pFET. Sub-threshold wing of less than 100mV/dec has been observed for both types of transistors even with relatively thick—12nm—gate oxide employed.

In the case of front channel conduction in nFET (Fig. 2(a)) sufficient back gate bias (V_{FG} =1 V) can turn on the conduction in the back channel. This can be seen more clearly when we have front channel transfer characteristics of nFET plotted with different back gate bias. In Fig. 3(a), threshold voltage change on a 300nm nFET can be maintained without turning on the back channel with back gate bias from -1 V to 0V. A similar study on the pFET has shown large threshold voltage change (-0.4 V to 1.2 V) was achievable. Therefore, individual transistor or even a series of back interconnected transistors can be dynamically set at either high V_T state for low leakage, or low V_T state for fast switching. Output currents can also be tuned for desired drive current in analog application. Such control of the device behavior offers the advantage of adaptive power control for

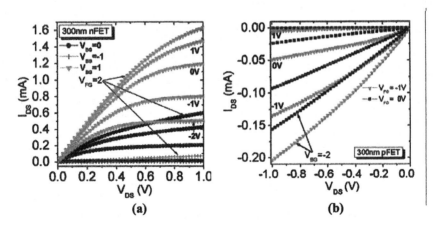

Fig. 1. (a) Front channel output characteristics of 300 nm nFET (L_{gate} = 300 nm; W_{gate} = 2 µm) and (b) back channel output characteristics of 300 nm pFET (L_{gate} = 300 nm; W_{gate}= 2 µm) at different front-gate and back-gate bias show good channel modulation from either gate.

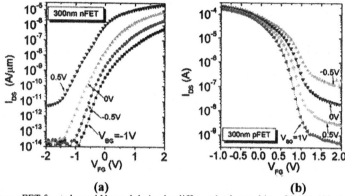

(a)   **(b)**

Fig. 3. (a) 300nm nFET front channel V_T modulation by different back-gate bias of 0.5 V, 0V, -0.5 V and -1V. (b) Back-gate-adjusted front channel transfer characteristics of a 300nm pFET incurring sufficient leakeage.

dynamic applications in both digital and analog circuit.

In addition, static current leakage reduction and better sub-threshold swing are also demonstrated, albeit on a leaky p-channel device (Fig. 3 (b)) with appropriate back-gate bias adjustment. With the back gate bias changed from 0 V to 0.5 V, leakage current has an improvement of 2 orders of magnitude (2×10^{-8} to 2×10^{-10}), and the sub-threshold swing move from above 100mV/dec to below that value. This indicates a valuable feature in improving performance robustness for speed, useful in logic and memory design.

4. Summary

Highly dense circuit implementation is made possible for adaptive power control with our buried interconnected back-gate devices structure. By carefully and dynamically setting the back gate bias, we can achieve low power, high speed and variable drive current for many logic and analog applications.

References

1. S. Tiwari, et. al., Electronics at nanoscale: fundamental and practical challenges, and emerging directions, *Emerging Technology-Nanoelectronics, IEEE Conference* 481-486, Singapore (2006).
2. T. Sekigawa and Y. Hayashi, *Solid-State Electronics.* 27, 827-828 (1984).
3. K. Kim, et. al., Nanoscale CMOS circuit leakage power reduction by double-gate device, *Digest of ISPLED International Symposium*, 102-107 (2004).
4. U. Avci, et. al., Back-gated SOI technology : power-adaptive logic and non-volatile memory using identical processing, *Digest of ESSDERC*, 285-288 (2004).
5. H. Lin, et. al., Super-Self-Aligned Back-Gate/Double-Gate Planar Transistors with thick Source/Drain and Thin Silicon Channel, *Digest of Device Research Conference*, 37-38, University Park (2006).
6. H-S. P. Wong, et. al., Nanoscale CMOS, *Proceedings of the IEEE*, 87, 537-570 (1999).

International Journal of High Speed Electronics and Systems
Vol. 17, No. 1 (2007) 147–152
© World Scientific Publishing Company

NON-VOLATILE HIGH SPEED & LOW POWER
CHARGE TRAPPING DEVICES

Moon Kyung Kim[*]

School of Electrical and Computer Engineering, Cornell University
Ithaca, NY 14853 U.S.A.
mkk23@cornell.edu

Sandip Tiwari

School of Electrical and Computer Engineering, Cornell University
Ithaca, NY 14853 U.S.A.
st222@cornell.edu

We report the operational characteristics of ultra-small-scaled SONOS (below 50 nm gate width and length) and SiO_2/SiO_2 structural devices with 0.5 um gate width and length where trapping occurs in a very narrow region. The experimental work summarizes the memory characteristics of retention time, endurance cycles, and speed in SONOS and SiO_2/SiO_2 structures. Silicon nitride has many defects to hold electrons as charge storage media in SONOS memory. Defects are also incorporated during growth and deposition in device processing. Our experiments show that the interface between two oxides, one grown and one deposited, provides a remarkable media for electron storage with a smaller gate stack and thus lower operating voltage. The exponential dependence of the time on the voltage is reflected in the characteristic energy. It is ~0.44 eV for the write process and ~0.47 eV for the erase process in SiO_2/SiO_2 structural device which is somewhat more efficient than those of SONOS structure memory.

Keywords: Nonvolatile memory, SONOS, Charge Trapping

1. Introduction

Many electronic applications require non-volatile memories (NVM) to retain information even when power is turned off or program the source of operating code that needs to be downloaded for the necessary computation. Obtaining ideal NVM characteristics such as low cost per bit, high density, high access speed, fast read/write cycle times, low power consumption, retention in excess of 10-years of retention time and 100K times of writing endurance is a key challenge in NVM electronics. However none of the current NVMs may satisfy all of these listed features at the smallest dimensions, and therefore there are always tradeoffs in real-life applications. Charge trapping memories have achieved most of these characteristics but the trade-offs between

[*] School of Electrical and Computer Engineering, 112 Phillips Hall, Cornell University, Ithaca, NY 14853

low power, fast read/write and retention have to be made. Two specific new approaches in NVM employ storage of single or few electrons such as in nano-crystal memories[1], and other modifications such as the use of silicon-oxide-nitride-oxide-silicon (SONOS) or back-floating gates[2,3]. All these approaches, in different ways, address the gate-stack thickness and voltage issues while attempting to achieve reproducibility and reliability. Nanocrystal memories utilize distributed and confined volume to store a few electrons, an approach that allows the charge injection oxide to be scaled while maintaining compatibility with mainstream silicon technology [4]. Silicon nitride and its interface with silicon dioxide provides an alternative for this charge storage where the highly localized storage of charge at increased number of sites may allow a further scaling of the insulator thickness and a trade-off in other attributes vis-à-vis nanocrystal memories. This report shows the memory characteristics of SONOS and SiO_2/SiO_2 structural devices and compares the characteristic energy of these devices which is strongly related to the retention time and trapping/detrapping time in charge trapping devices.

2. Experiments

Figure 1 shows the fabrication sequence employed for the small SONOS memory structures with SOI substrates. Mixed-mode lithography, combining optical and electron-beam, is used to obtain the nanoscale SONOS device. For the SiO_2/SiO_2 structural device, only optical lithography is employed to get a 0.5 um scale structure.

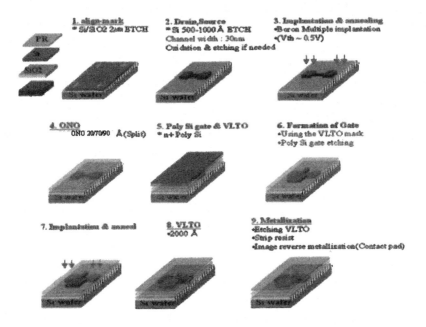

Fig. 1. SONOS memory process flow based on mixed-lithography.

Following the active region definition on the thinned silicon-on-oxide using sacrificial dielectric stack, the oxide-nitride-oxide (2 nm/7 nm/ 9 nm) and the oxide-oxide stack (3 nm/7 nm) are grown and deposited. The bottom tunnel oxide is grown through thermal oxidation and then nitride stack and the blocking oxide are deposited by low-pressure chemical vapor deposition (LPCVD). These nitride layer in SONOS and two different oxides in SiO_2/SiO_2 structural device work as trap sites to store electrons. Polysilicon gate patterning and sidewall process are utilized in making the devices. Figure 2 shows

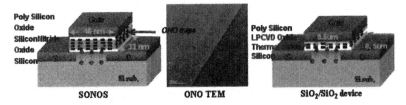

| SONOS | ONO TEM | SiO_2/SiO_2 device |

Fig. 2. Schematics of SONOS & SiO_2/SiO_2 structural device and TEM image of ONO

the schematic diagrams of SONOS and SiO_2/SiO_2 device together with a transmission-electron micrograph (TEM) of the cross-section of this grown and deposited memory stack with the dark region as the silicon nitride.

3. Results and Discussion

Figure 3 shows scanning electron micrograph (SEM) of the smallest dimensional SONOS device and micrometer scale SiO_2/SiO_2 structural device respectively.

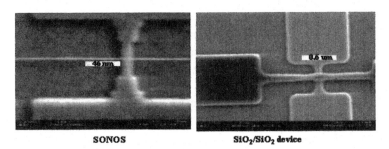

| SONOS | SiO_2/SiO_2 device |

Fig. 3. SEM images of SONOS and SiO_2/SiO_2 structural device

Figure 4 shows the writing and erasing characteristics at various voltages for a SONOS and SiO_2/SiO_2 structural device. Very thin gate stacks improve program/erase characteristics. To achieve 1.5 V threshold voltage shift in SONOS and SiO_2/SiO_2 structural device, 13 V pulse is applied to the gate for 200 usec and 500 usec respectively. In the erasure process, 5 msec for SONOS and 1 msec for SiO_2/SiO_2 structural device are required at -13 voltage to obtain -1.5 V threshold voltage shift. The reduction of the effective barrier by applying larger write and erase voltages causes the higher tunneling probabilities. This is observed in these plots of figure 4(a) and 4(b) as a decrease in

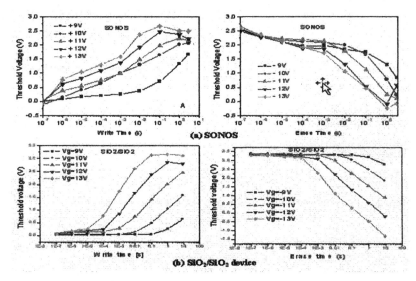

Fig. 4. Write and erase characteristics of (a) SONOS and (b) SiO₂/SiO₂ structural device

writing and erasing times with increased applied tunneling voltages. The capture process is based on Fowler-Nordheim injection and the erasure process is presumably a Poole-Frenkel mechanism, or some other similar detrapping process with strong localization and field-dependence. This different mechanism can be considered by looking at the characteristic energy (E_{ch}) of the capture and emission processes using the time-dependence of the response to write or erase voltages (Fig. 4). This characteristic energy is the Coulomb charging energy and defects such as point-like defects, voids, dislocations, impurities etc. have coulomb charging energy comparable to or smaller than the semiconductor bandgap. Nearly all of such tunneling-based injection processes that have been modeled have a characteristic bias/energy dependence that is inversely exponentially related to the thickness and barrier height. The transmission rate is

$$T_t = C \exp\left[-\frac{E_{ch}}{qV}\right], \tag{1}$$

where C is reduced due to localization and reduced coupling of defect and has a field dependence, but E_{ch} still captures the effect of the barrier. This dependence allows us to write the capture and emission time (t) for the change in the threshold voltage of ΔV_T as:

$$t = \beta \, \Delta V_T \exp\left(\frac{E_{ch}}{qV}\right), \tag{2}$$

where we have derived the time through integration of injected charge trap proportional to the current density. The exponential dependence of the time on the voltage is reflected in the characteristic energy (Ech) which is ~0.44 eV for the write process and ~0.47 eV for the erase process in SiO2/SiO2 structural device. SONOS has a higher characteristic

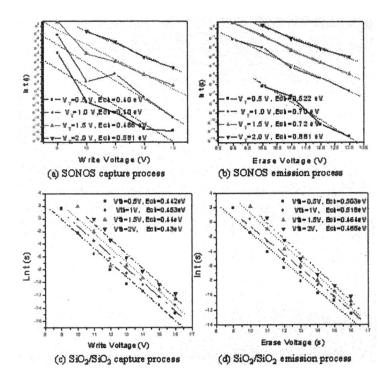

Fig. 5. Capture/emission characteristic energy of SONOS and SiO₂/SiO₂ structural device

energy of 0.49 eV for capture process and 0.72 eV for emission process. The writing process is somewhat more efficient than the erasure process as can be seen in figure 4 and figure 5. This lower characteristic energy in SiO_2/SiO_2 structural device causes negative effects on retention time when compared with SONOS structures. Figure 6 shows the retention and endurance characteristics in SiO_2/SiO_2 structural device.

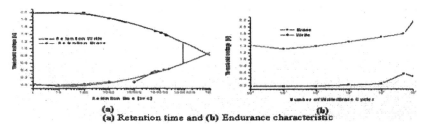

(a) Retention time and (b) Endurance characteristic

Fig.6. Capture/emission characteristic energy of SONOS and SiO₂/SiO₂ structural device

As mentioned for the characteristic energy, retention characteristics are poor. However, the endurance is over 10^5 cycles keeping the initial memory window in SiO_2/SiO_2 structural device. SONOS satisfies the retention and endurance characteristics needed. Retention time can be improved by optimizing the blocking oxide thickness or by making deep level-energy trap sites.

4. Summary

In conclusion, SONOS and SiO_2/SiO_2 charge trapping structural device have been characterized and the devices show well-functioned trapping memory. The structures do have attractive endurance and fast programming/erasing properties. Lower characteristic energy in SiO_2/SiO_2 structural device causes poorer retention compared to SONOS device. This study shows that SONOS and SiO_2/SiO_2 charge trapping structural device might be the good candidates for the future NVMs. The retention time in SiO_2/SiO_2 structural device needs further improvement.

5. Acknowledgments

This work was supported by NSF through Cornell Center for Materials Research and the National Program for Tera level Nano Devices through MOST, and was performed at Cornell Nanoscale Facility.

References

1. S. Tiwari et al., "A silicon nanocrystals based memory", Appl. Phys. Lett. 68, pp.1377-1379, (1996)
2. F. R. Libsch and M. H. White, "Charge Transport and Storage of Low Programming Voltage SONOS/MONOS memory Devices", Solid State Electronics, Vol.33, No.1, pp.105-126, (1990)
3. H. Silva, Moon Kyung Kim, Prof.S.Tiwari "Scaled Front-Side and Back-Side Trapping SONOS Memories on SOI", SOI Conference Proceedings (2003)
4. MoonKyung Kim, S. Tiwari, "Ultra-short SONOS memories",IEEE Tran. on nanotechnology,v3, 4, 417- 424 (2004)

International Journal of High Speed Electronics and Systems
Vol. 17, No. 1 (2007) 153–162

HIGH PERFORMANCE SIGEC/SI NEAR-IR ELECTROOPTIC MODULATORS AND PHOTODETECTORS

MARTIN SCHUBERT

Electrical and Computer Engineering, Cornell University,
Ithaca, NY 14853, United States
mfs24@cornell.edu

FARHAN RANA

Electrical and Computer Engineering, Cornell University,
Ithaca, NY 14853, United States
fr37@cornell.edu

Numerical simulations are performed for SiGeC/Si electrooptic modulators and photodetectors operating at near-IR wavelengths. The addition of carbon provides the ability to lattice match layers with high germanium composition to silicon, which is shown to allow structures with a substantial increase in the optical confinement factor. In addition, SiGeC/Si heterostructures provide strong confinement of large electron and hole concentrations. The large optical confinement factor and strong carrier confinement enable broadband electrooptic modulators with sub-100 μm lengths and switching times below 0.5 ns with 25 mA current as well as photodetectors with quantum efficiencies as high as 90% for 300 μm length.

Keywords: Device modeling, integrated optics, photodetector, optical modulator, plasma dispersion effect, silicon optoelectronics.

1. Introduction

The advantages of CMOS compatible manufacturing have made the silicon material system an attractive target for photonic system applications. However, both all-silicon and SiGe/Si devices suffer from fundamental limitations that hinder performance and system integration. An all-silicon platform does not allow for both light modulation and photodetection at the same wavelength. In addition, the silicon-on-insulator (SOI) designs which are typically used have poor thermal characteristics which keep maximum current low and limit device speed or bandwidth [1]-[3]. Finally, it is difficult to design structures that confine both the electrons/holes and photons to the same location in space while keeping the optical mode away from highly doped contact layers [1]-[3]. Although SiGe/Si heterostructures have the potential to meet these challenges, the large lattice mismatch between germanium and silicon results in very small values for the critical thickness of epitaxial layers. Consequently, the results for SiGe modulators and detectors that have been reported in literature suffer from small optical confinement factors, long device lengths, and small RC limited bandwidths [4]-[10].

Silicon-germanium-carbon structures make it possible to tightly confine both carriers and the optical mode to the same spatial location, thereby offering design opportunities similar to those available in optoelectronic devices based on III-V materials. SiGeC has been used previously in optoelectronics for waveguiding [11] and photodetectors [12]-[14]. In this work, thick lattice-matched layers with high germanium content are used as the core for optical waveguides to achieve high modal confinement factors. The high confinement factors achievable with SiGeC/Si heterostructures combined with strong electron/hole confinement and superior thermal characteristics can produce integrated electro-optic modulators and photodetectors with significant performance benefits over conventional designs.

2. Material Properties

Typically up to 4% carbon can be incorporated in SiGeC epitaxial layers, which allows for thick defect-free layers with high germanium content that are perfectly lattice matched to silicon [15]. Fig. 1 shows the Matthews-Blakeslee critical thickness of SiGeC layers as a function of the composition. Germanium content over 30% can be reached in lattice-matched layers with carbon content below the 4% limit.

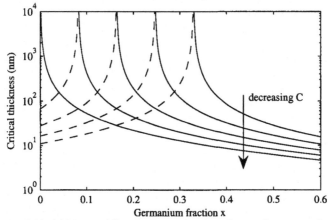

Fig. 1 – Critical thickness of $Si_{1-x-y}Ge_xC_y$ as a function of germanium content for carbon fractions of 0.04, 0.03, 0.02, 0.01, and 0. Solid curves are for alloys which are compressively strained, dashed lines are for tensile strain.

The refractive index model used for SiGeC depends only on the bandgap and the indexes of bulk silicon and germanium. The refractive index of a SiGeC alloy is taken to be that of a SiGe alloy with the same bandgap. Using a linear model for the refractive index of SiGe from [5] and expressions for the bandgap of SiGe and SiGeC from [15] yields,

$$n_{Si_{1-x-y}Ge_xC_y} = n_{Si} + \left(n_{Ge} - n_{Si}\right)\left(x - \frac{2.1}{0.9}y\right) \qquad (1)$$

The plasma dispersion effect and free carrier dispersion are modeled as identical to silicon as given in [16].

High germanium content in the core layer can lead to large material absorption at near-IR wavelengths. In this paper, material absorption is calculated using a model for phonon-assisted indirect optical transitions similar to one used previously for SiGe [5],

[17]. The effect of carbon and strain is neglected with the exception of the modification of the bandgap. The addition of carbon to SiGe has two effects on the alloy bandgap. Small carbon atoms relax the strain in SiGe and allow the bandgap to increase. The intrinsic effect of adding carbon actually decreases the bandgap because of a large bowing parameter. The net effect of adding carbon is a slight increase in the bandgap, which leads to a slight decrease in the absorption coefficient. Fig. 2 plots the absorption coefficient as a function of composition.

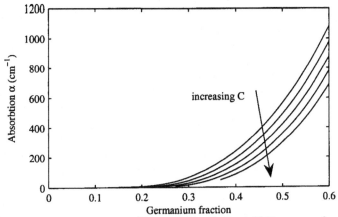

Fig. 2 – Optical absorption coefficient of $Si_{1-x-y}Ge_xC_y$ at 1300 nm as a function of germanium content for carbon fractions of 0, 0.01, 0.02, 0.03, and 0.04.

3. Electrooptic Modulator Design

In simulations we consider p-i-n diode modulators in the familiar Mach-Zehnder configuration operating at 1550 nm. The phaseshift due to a change in index in the active region is given by,

$$\Delta\phi = \frac{2\pi}{\lambda}\Gamma\Delta n_{act}L \tag{2}$$

where Γ is the optical confinement factor and L is the modulator length. The modulator switches from the on-state to the off-state when $\Delta\phi = \pi$. In silicon the index change due to the presence of carriers is approximately linearly related to the change in carrier concentration [16]. With a current injection level I and carrier recombination and leakage neglected, this yields

$$\Delta\phi \approx \frac{2\pi}{\lambda}\frac{\Gamma}{hw}\frac{1}{q}fIt \tag{3}$$

where h and w are the active region height and width, t is time, and f relates the carrier concentration to the change in index. The key figure of merit for modulator designs is Γ/hw – maximizing this term gives the fastest time to achieve a π phaseshift with a given current. Simulations performed assumed a width of 0.5 μm which gives nearly complete lateral confinement. Fig. 3 and 4 plot Γ and Γ/h for a modulator at 1550 nm as a function of active region height for several carbon fractions. At each thickness, the germanium composition is maximized while keeping within the limits of the critical thickness. Adding carbon allows a higher germanium composition for a given thickness and yields a higher refractive index. As a result, the confinement factor for a given thickness goes up and Γ/*h* increases by up to a factor of 5. Optimized designs for a given carbon concentration are listed in table 1, and will be used later when treating the modulator electrical characteristics.

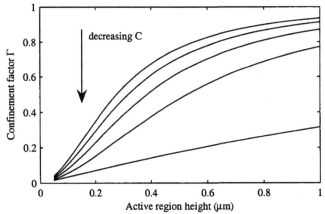

Fig. 3 – Active region confinement factor Γ as a function of active region height for modulators with carbon compositions of 0.04, 0.03, 0.02, 0.01, and 0.

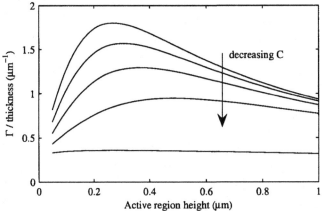

Fig. 4 – Figure of merit Γ/*h* as a function of active region height for the modulator with carbon composition y of 0.04, 0.03, 0.02, 0.01, and 0.

% carbon	% germanium	core height (μm)	Γ	Γ/h (μm⁻¹)
0.0	2.51	0.28	0.10	0.36
1.0	9.84	0.48	0.45	0.95
2.0	18.5	0.37	0.48	1.30
3.0	27.1	0.31	0.49	1.57
4.0	35.6	0.27	0.49	1.80

Table 1 – Parameters for optimized modulator designs.

Carrier injection into the modulator is modeled using the carrier rate equation,

$$\frac{\partial N}{\partial t} = \frac{I}{qV} - \frac{N}{\tau_{rec}} - \frac{N}{\tau_{leak}(N)} \tag{4}$$

where V is the total core region volume, τ_{rec} is the carrier recombination lifetime, and $\tau_{leak}(N)$ is the carrier density dependent time constant associated with current leakage out of the active region due to thermionic emission. In simulations τ_{rec} was assumed to be 20 ns, while values for τ_{leak} were extracted from simulations using the semiconductor simulation package SimWindows [18]. As injected carrier concentrations grow, the relative fraction of current which leaks from the active region grows larger and τ_{leak} decreases. Fig. 5 plots the carrier leakage time constant as a function of the injected carrier density for the optimized devices. Modulators with large germanium and carbon concentrations have relatively large band offsets which aid in confining carriers, resulting in values for τ_{leak} that are larger by a factor of nearly 10^4.

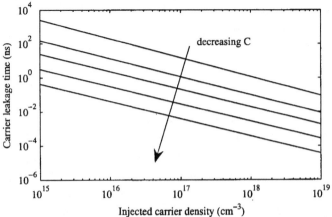

Fig. 5 – Carrier leakage lifetime as a function of carrier concentration for optimized modulators with carbon content of 0.04, 0.03, 0.02, 0.01, and 0. The doping is given by $N_d = 1.5 \times 10^{18}$ and $N_a = 3.0 \times 10^{18}$.

Solving eq. (4) for the carrier density which gives a π phase shift gives the turn-on time for the modulator. The turn-on time for a modulator with 500 μm length is plotted in fig. 6 for the optimized modulator designs. Decreasing the modulator length requires higher current injection and makes the carrier leakage more severe. For a fixed current, as the length becomes smaller the modulator may no longer be able to achieve a π phase shift. Fig. 7 plots the turn-on time of optimized modulators with a fixed 25 mA current level as a function of the device length. Modulators with the highest carbon content achieve sub-ns switching times with lengths approaching 10 μm. At these lengths the RC characteristics should not be a limitation for device performance. For the shortest devices, heating will hinder device operation. However, device heating due to large current densities is expected to be less of a problem with SiGeC/Si modulators compared to SOI modulators because of the absence of the poor-thermal-conductivity oxide layer separating the device from the substrate. Detailed analysis of modulator performance including thermal effects is left for later work.

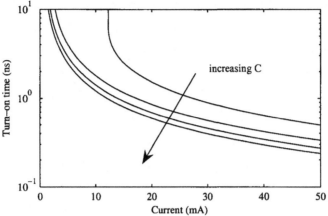

Fig. 6 – Turn-on time as a function of current for optimized modulators with carbon fractions of 0.01, 0.02, 0.03, and 0.04. The device length is 500 μm.

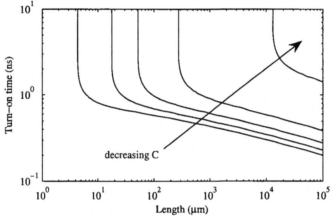

Fig. 7 – Turn-on time of optimized modulators for a 25 mA current as a function of the modulator length with carbon fractions of 0.04, 0.03, 0.02, 0.01, and 0.

The turn-on time is typically the dominant component of switching speed. Turn-off time is the time needed to remove the carriers from the active region, which is accomplished by reverse biasing the device and can be fast [1]. Simulated carrier sweep-out times in [1] are less than 0.2 ns for a SOI modulator with a 0.5 μm wide junction. Devices considered here have junction widths that are less than 0.5 μm and have a device geometry that generates strong well confined electric fields in the core layer. As a result we expect the turn-off time will be less than for SOI devices and the speed of the modulator is well predicted by calculations for the turn-on time in figs. 6 and 7.

4. Photodetector Design

The addition of carbon to SiGe allows larger critical thicknesses for high germanium content layers and enables the use of SiGeC separate-confining-heterostructures (SCHs) which increases the confinement factor of the optical mode in the active region. Initially we consider photodetectors at 1300 nm with a single active region surrounded on both sides by an SCH. Fig. 8 plots the confinement factor as a function of the germanium composition for the photodetectors with several carbon fractions. For each curve, the composition of the active region is given by indicated germanium and carbon fractions. The carbon fraction of the SCH is the same as the active region, while the germanium fraction is that which lattice matches the layer to the silicon substrate. The SCH thickness is optimized at each point to maximize the confinement factor. Raising the carbon composition increases the confinement factor because the critical thickness for the germanium-rich active region is increased, and the refractive index of the SCH also rises. Fig. 9 plots the modal absorption for the waveguide photodetectors as a function of composition. While the material absorption of SiGeC actually decreases as carbon is added (see fig. 2), the modal absorption of optimized modulators grows with the addition of carbon. This is because the critical thickness increases rapidly enough to overcome the reduction in material absorption. The net effect is a strong increase in absorption.

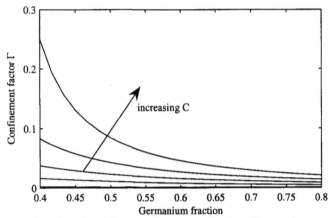

Fig. 8 – Optical mode confinement factor as a function of germanium content for photodetectors with carbon content of 0, 0.01, 0.02, 0.03, and 0.04.

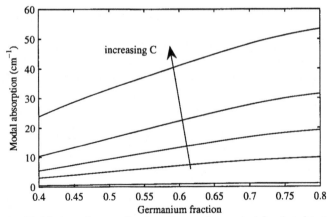

Fig. 9 – Modal absorption as a function of germanium content for photodetectors with carbon content of 0, 0.01, 0.02, 0.03, and 0.04.

Superlattice active regions can also improve the modal overlap with germanium-rich active layers. Fig. 10 shows absorption as a function of the number of quantum wells for several carbon fractions and a germanium composition of 0.6. Maximum performance is achieved for devices with two periods. Increasing the number of superlattice periods further forces the absorbing sections to become narrower or to grow farther apart – both of which decrease the confinement factor and modal absorption. Fig. 11 plots the quantum efficiency of optimized photodetectors 300 μm in length. Values as high as 90% are achieved for the photodetector with maximum carbon content, which compares favorably with a value of 12% for a 1 mm SiGe/Si device in [5] using a similar critical thickness model.

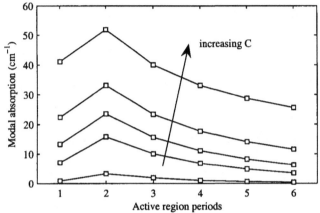

Fig. 10 – Modal absorption as a function of the number of active region superlattice periods

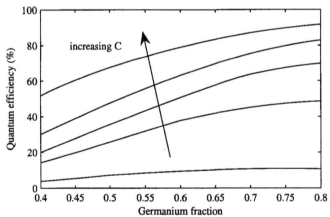

Fig. 11 – Turn-on time of optimized modulators for a 25 mA current as a function of the modulator length with carbon fractions of 0.04, 0.03, 0.02, 0.01, and 0.

5. Conclusion

We have proposed and analyzed broadband electrooptic modulators and photodetectors based upon the SiGeC/Si material system. SiGeC can be lattice-matched to silicon, which allows for thick layers with large germanium content that can be used to achieve high optical mode confinement in optical waveguides. Additionally, the band offset between Si and SiGeC strongly confines carriers to the waveguide core, which results strong overlap between large injected carrier concentrations and the optical mode. Modulators which take advantage of these properties can be designed with lengths less than 100 μm and switching times below 0.5 ns, while photodetectors can be made with efficiencies as high as 90% for 300 μm lengths.

References

1. C. A. Barrios, V. R. Almeida, R. Panepucci, and M. Lipson, "Electrooptic Modulation of Silicon-on-Insulator Submicrometer-Size Waveguide Devices," *J. Lightwave. Tech.*, vol. 21, no. 10, p. 2332, (2003)
2. C. A. Barrioes, V. Almeida, M. Lipson, "Low-Power-Consumption Short-Length and High-Modulation-Depth Silicon Electrooptic Modulator," *IEEE J. Lightwave Technol.*, 21, p. 1089 (2003)
3. A. Liu, R. Jones, L. Liao, D. Samaro-Rubio, D. Rubin, O. Cohen, R. Nicolaescu, M. Paniccia, "A high-speed silicon optical modulator based on a metal-oxide-semiconductor capacitor," *Nature*, 427, p. 615 (2004)
4. B. Li, Z. Jiang, X. Zhang, X. Wang, J. Wan, G. Li, E. Liu, "SiGe/Si Mach-Zehnder interferometer modulator based on the plasma dispersion effect," *Appl. Phys. Lett.*, 74, p. 2108 (1999).
5. L. Naval, B. Jalali, L. Gomelsky, J. M. Liu, "Optimization of $Si_{1-x}Ge_x$/Si Waveguide Photodetectors Operating at $\lambda = 1.3$ μm" *J. Lightwave Technol.*, 14, p. 787 (1996)
6. B. Jalali, A. F. J. Levi, F. Ross, E. A. Fitzgerald, "SiGe waveguide photodetectors grown by rapid thermal chemical vapor deposition," *Electronics Letts.*, 28, p. 269 (1992)
7. A. Splett, T. Zinke, E. Kasper, H. Kibbel, H. J. Herzog, H. Presting, "Integration of waveguides and photodetectors in SiGe for 1.3 μm operation," *IEEE Photonics Techonl. Letts.*, 6, p. 59 (1994)
8. H. Temkin, T. P. Pearsall, J. C. Bean, R. A. logan, S. Luryi, "Ge_xSi_{1-x} strained-layer superlattice waveguide photodetectors operating near 1.3 μm" *Appl. Phys. Letts.*, 48, p. 963 (1986)
9. T. Tashiro, T. Tatsumi, M. Sugiyama, T. Hashimoto, T. Morikawa, "A selective epitaxial SiGe/Si planar photodetector for Si-based OEIC's," *IEEE Trans. Electron Dev.*, 44, p. 545 (1997).
10. D. Marries, E. Cassan, L. Vivien, D. Pascal, A. Koster, and S. Laval, "Design of a modulation-doped SiGe/Si optical modulator integrated ina submicrometer silicon-on-insulator waveguide," *Optical Engineering*, vol. 4(8), p. 084001-2, 2005
11. R. A. Soref, Z. Atzman, F. Shaapur, M. Robinson, R. Westhoff, "Infraraed waveguiding in SiGeC upon Silicon," Optics Letts., vol. 21, no. 5, p. 345, 1996.
12. F. Y. Huang and K. L. Wang, "Normal-incidence epitaxial SiGeC photodetector near 1.3 μm wavelength grown on Si substrate," *Appl. Phys. Lett.*, vol. 69, no. 16, p. 2330, (1996)
13. F. Y. Huang, K. Sakamoto, K. L. Wang, P. Trinh, and B. Jalali, "Epitaxial SiGeC Waveguide Photodetector Grown on Si Substrate with Response in the 1.3–1.55-μm Wavelength Range," *IEEE Photon. Technol. Lett.*, vol. 9, no. 2, p. 229, (1997)
14. B. Li, S.-J. Chua, E. A. Fitzgerald, "Theoretical analysis of $Si_{1-x-y}Ge_xC_y$ near-infrared photodetectors," *Optical Engineering*, vol 42, no. 7, p. 1993, (2003)

15. S. T. Pantellides, S. Zollner, *Silicon-Germanium-Carbon Alloys*, Taylor and Francis, NY (2002)
16. R. A. Soref and B. R. Bennett, "Electrooptical Effects in silicon," *IEEE J. Quantum Electron.*, vol. QE-23, no. 1, p. 123, 1987
17. R. Braunstein, A. R. Moore, F. Herman, "Intrinsic Optical Absorption in Germanium-Silicon Alloys," *Physical Review*, 109, p. 695 (1958)
18. D. W. Winston, "Physical simulation of optoelectronic semiconductor devices," Ph.D. thesis, Univ. of Colorado at Boulder, 1996

Section IV.
Nanoelectronics and Ballistic Devices

International Journal of High Speed Electronics and Systems
Vol. 17, No. 1 (2007) 165–172
© World Scientific Publishing Company

HYBRID NANOMATERIALS FOR MULTI-SPECTRAL INFRARED PHOTODETECTION

ADRIENNE D. STIFF-ROBERTS

Department of Electrical and Computer Engineering, Duke University, Box 90291,
Durham, North Carolina 27708-0291, USA
adrienne.stiffroberts@duke.edu

Quantum dot infrared photodetectors (QDIPs) using quantum dots (QDs) grown by strained-layer epitaxy have demonstrated low dark current, multi-spectral response, high operating temperature, and infrared (IR) imaging. However, achieving near room-temperature, multi-spectral operation is a challenge due to randomness in QD properties. The ability to control dopant incorporation is important since charge carrier occupation influences dark current and IR spectral response. In this work, dopant incorporation is investigated in two classes of QDs; epitaxial InAs/GaAs QDs and CdSe colloidal QDs (CQDs) embedded in MEH-PPV conducting polymers. The long-term goal of this work is to combine these hybrid nanomaterials in a single device heterostructure to enable multi-spectral IR photodetection. Two important results towards this goal are discussed. First, by temperature-dependent dark current-voltage and polarization-dependent Fourier transform IR spectroscopy measurements in InAs/GaAs QDIPs featuring different doping schemes, we have provided experimental evidence for the important contribution of thermally-activated, defect-assisted, sequential resonant tunneling. Second, the enhanced quantum confinement and electron localization in the conduction band of CdSe/MEH-PPV nanocomposites enable intraband transitions in the mid-IR at room temperature. Further, by controlling the semiconductor substrate material, doping type, and doping level on which these nanocomposites are deposited, the intraband IR response can be tuned.

Keywords: Epitaxial quantum dots; colloidal quantum dots; hybrid nanocomposites, infrared photodetection.

1. Achieving Multi-spectral Infrared Photodetection Using Hybrid Nanomaterials

The detection of IR radiation is critical for many applications, such as situational awareness sensors and thermal imaging. Commercial technologies, such as Si microbolometers and HgCdTe narrow-bandgap photodiodes, are widely used; yet, significant improvements in device performance can be realized by using a QD active region. In fact, InAs/GaAs QDIPs, which use intraband transitions, are rapidly approaching establishment as a viable alternative in the mid-IR, having demonstrated low dark current, multi-spectral response, high-detectivity, high operating temperature, and

IR imaging[1-3]. In particular, the low dark current resulting from three-dimensional quantum confinement enables high-operating temperature IR photodetection ($\geq$ 150 K).

InAs/GaAs QDs are synthesized using strained-layer epitaxy in ultra-high vacuum crystal growth systems, such as molecular beam epitaxy (MBE). QD growth is a consequence of the lattice mismatch between the wider-bandgap matrix material and the narrower-bandgap dot material, and growth results in the formation of three-dimensional islands by the minimization of strain energy. Depending on the band-lineup of the dot and matrix materials, confinement barriers are created in the conduction and/or valence bands, thereby enabling independent control of electron and/or hole populations, respectively, through intraband transitions[4]. Analogous intraband transitions could be achieved using colloidal quantum dots (CQDs)[5-7] embedded in conducting polymers, such as poly[2-methoxy-5-(2-ethylhexyloxy)-1,4-phenylenevinylene] (MEH-PPV). In this way, the conducting polymer provides quantum confinement barriers. CQD active regions embedded in such a polymer could significantly improve QDIP performance due to: i) the ability to yield highly-uniform ensembles of nanostructures through size-filtering, and ii) the simplification of device design since quantum-sized effects are related to spherical CQDs. In addition, CQDs embedded in a semiconductor matrix would enable a greater selection of active region materials for optoelectronic devices since the chemical synthesis of CQDs eliminates strain considerations. It is important to note that the wavefunctions and corresponding energy levels for charge carriers quantum-confined in CQD/polymer nanocomposites has been reported[8], thereby providing a theoretical foundation for the observation of these intraband transitions. Further, the CdSe CQD/MEH-PPV polymer nanocomposite material system is well-suited for the observation of mid- and long-wave IR, intraband transitions due to the ~1.5 eV difference between their electron affinities, which corresponds to the conduction band offset for CdSe and MEH-PPV. Therefore, CdSe CQDs embedded in MEH-PPV provide electron quantum confinement such that intraband transitions can occur in the conduction band.

The long-term goal of this work is to combine both classes of QDs in a single device heterostructure to achieve multi-spectral IR photodetection. The impact of using such hybrid nanomaterials in a monolithic, semiconductor-based heterostructure is the compact, integrated, and rugged packaging of a complex system on a chip, such as a gas-sensing spectrometer enabled by hyperspectral detection in the infrared range. In this work, the following topics are presented: 1) dark current voltage and Fourier transform IR (FTIR) absorbance measurements of InAs/GaAs QDIPs featuring different doping schemes (modulation- vs. delta-doping) and doping concentrations (1, 2, and 3 electrons/dot) to provide a comprehensive picture of dopant incorporation for device optimization; and 2) FTIR absorbance measurements of CdSe/MEH-PPV nanocomposites deposited on semiconductor substrates (Si, Ge, and GaAs featuring different doping types and concentrations) to demonstrate a method for controlling charge transfer in nanocomposites to achieve intraband, IR absorption.

2. Dopant Control in InAs/GaAs Quantum Dot Infrared Photodetectors

The fundamental challenge in InAs/GaAs QDIPs is the non-uniformity of QD ensembles grown by strained-layer epitaxy. The resultant inhomogeneous linewidth broadening and limited control of energy levels precludes the reduction of dark current by three-dimensional quantum confinement. It should be possible to decrease dark current density and to improve spectral response control with better understanding of how dopants are incorporated into high-density QD ensemble energy levels. Through the investigation of device heterostructures with varying modulation- and delta-doped carrier concentrations, we have correlated the charge filling process of energy levels in high-density QD ensembles with temperature-dependent dark current-voltage (and thereby the thermal activation energy) and FTIR absorbance measurements.

Fig. 1 shows a schematic diagram of the thirty-layer, InAs/GaAs QDIP heterostructures grown in this dopant incorporation study. Six samples were grown by solid source MBE using a Riber 2300 system at Duke University. Three samples featured modulation-doping with nominal carrier densities of 1, 2, and 3 electron(s)/dot, while three samples featured delta-doping with the same carrier densities. Each QDIP sample was grown on a semi-insulating (100) GaAs substrate with background pressure of 4×10^{-11} Torr and a cracked As_2 source. It is important to note that modulation-doping is distinguished from delta-doping by the doped GaAs layer thickness (4nm and 0 nm,

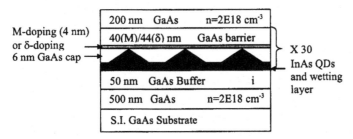

Fig. 1. Schematic diagram of the InAs/GaAs QDIP device heterostructure cross section.

respectively). The GaAs top and bottom contact layers, as well as the GaAs buffer layer, were grown at 620 °C, while the InAs QD and GaAs barrier layers were grown at 500 °C. The formation of 2.2 ML InAs QDs was monitored by reflection high-energy electron diffraction, and a 30 sec pause allowed complete QD formation. The typical characteristics of the QD layers include surface density $\sim 1 \times 10^{11} cm^{-2}$, average height ~6nm, average base width ~25-30nm, and interdot spacing ~4-7nm. It is important to note that the doping location used in these samples was determined from a previous study in which QDIP performance was optimized as a function of dopant location at 78K[9].

Fig. 2(a) shows the thermal activation energies calculated from temperature-dependent (17-300K), dark current-voltage measurements as a function of average electron occupation per QD for both modulation- and delta-doped samples. Note that delta-doped samples have lower activation energies than modulation-doped samples. In addition, the activation energies are all < 30 meV, which is much smaller than the expected activation energy to thermally excite a ground-state electron out of the QD to the continuum (~100 meV). These small activation energies are indicative of thermally-activated, defect-assisted sequential resonant tunneling from QDs, and this transport mechanism contributes significantly to dark current in InAs/GaAs QDIPs. Fig. 2(b) shows representative polarization-dependent FTIR absorbance peaks for an InAs/GaAs QDIP modulation-doped with 2 electrons per QD at 150K. It is important to note that in addition to the QD absorbance peaks near 90 meV, an additional peak at 405 meV (that does not vary with polarization) is observed. This energy is close to the thermal emission activation energy of DX centers in Si-doped GaAs (~330meV)[10-12]. The 405meV peak is most likely associated with the optical activation energy of the DX center. Thus, these observed DX centers could be the source of sequential resonant tunneling in QDIPs, and understanding how they interact with InAs/GaAs QDs should enable better control of dopant incorporation and reduced dark current for higher temperature operation.

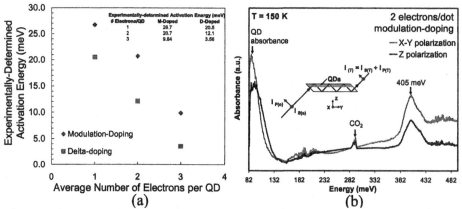

Fig. 2. (a) Thermal activation energies calculated from Arrhenius plots of dark current-voltage measurements ranging from 17-300K for InAs/GaAs QDIPs featuring different doping techniques and concentrations; and (b) X-Y- and Z-polarized FTIR absorbance spectra for modulation-doped concentration of 2 electrons/dot at T = 150K. The inset shows a schematic of the sample waveguide geometry for polarization-dependent measurements.

3. Fourier Transform Infrared Absorbance in CdSe/MEH-PPV Nanocomposites on Semiconductor Substrates

Interband transitions in II-VI CQDs typically fall within the visible range, yet it is desirable to investigate the application of CQDs to the IR regime. Some initial work has already demonstrated IR sensitivity in CQDs, such as electroluminescence[13,14] and photoconductivity[15] in the near-IR (1-3 μm). Typically, these near-IR transitions are observed using pump-probe, photo-induced (visible) absorption measurements and are due to intraband absorption from the $1S_e$ to $1P_e$ electronic levels of excitonic transitions[16-18]. In addition, these intraband transitions are usually observed in dispersed CQDs or CQD solids featuring matrix materials that do not provide additional quantum confinement. Also, interband transitions in the mid-IR have been demonstrated using PbS and PbSe CQDs[19-21]. The unique approach of this work is that MEH-PPV is used to provide an additional confinement barrier to the conduction band of CdSe CQDs. Thus, while the bandgaps of CdSe and MEH-PPV correspond to visible light, a wide range of mid-IR absorbance peaks are observed using FTIR absorbance spectroscopy.

In this study, CdSe CQD/MEH-PPV polymer nanocomposites (~ 30% wt.) dissolved in toluene were drop-cast at room-temperature on a series of semiconductor substrates, including semi-insulating (SI) Si, Ge, and GaAs, as well as a range of p- and n-type doped GaAs. The CdSe CQDs and MEH-PPV polymer were purchased commercially from NN-Labs and American Dye Source, Inc., respectively, while the resultant nanocomposites were synthesized at Duke University. The SI semiconductor substrates were also purchased from commercial providers, while the p- and n-type GaAs substrates were grown using the Riber 2300 MBE system at Duke (0.3 μm bulk GaAs epitaxial layers grown on semi-insulating GaAs substrates). Each substrate was cleaned using buffered HF acid, acetone, and isopropyl alcohol before nanocomposite deposition. The control over the doping of the GaAs substrate is important in controlling the Fermi energy, which impacts the CQD carrier occupation, as shown in the following sections. The CdSe CQDs range in size from ~ 6.2 to 7.7 nm. The surface ligands surrounding these CQDs are octadecylamine (ODA), and the bandgap energy is ~1.94 eV. The conducting polymer, MEH-PPV, features a bandgap energy of ~2.28 eV and a Fermi energy approximately 0.5 eV above the valence band, thereby behaving like a p-type semiconductor. It is also important to note that reference samples were prepared by depositing MEH-PPV on each substrate in order to extract the CdSe CQD IR absorbance.

The results from this experiment are summarized generally in Figures 3(a) and (b), which show the FTIR absorbance spectra on SI substrates and the FTIR absorbance peak variation with GaAs doping concentration, respectively, for the CdSe CQD/MEH-PPV polymer nanocomposites. It is important to note that these mid-IR peaks are all observed at room-temperature. In addition, as shown in Fig. 3(b), the IR peak is blue-shifted with increasing p-type doping concentration and is red-shifted with increasing n-type doping concentration.

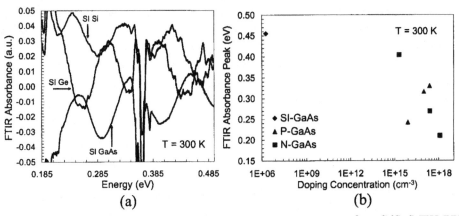

Fig. 3. (a) FT-IR absorbance spectra at room-temperature for CdSe/MEH-PPV nanocomposites deposited on semi-insulating Si, Ge, and GaAs substrates; and (b)room-temperature, FTIR absorbance peak variation with doping concentration for SI, p-type, and n-type GaAs substrates.

As seen in Figure 4, the doping characteristics of the semiconductor substrate affect the heterojunction interface band-bending and provide a means by which the IR absorbance peak can be tuned. For example, depending on the substrate doping type (semi-insulating, p-type, or n-type) different IR absorbance peaks become dominant in the CdSe CQD. While the Fermi energies of semi-insulating GaAs and MEH-PPV are close and very little band-bending occurs at their interface, p-type GaAs provides an electron accumulation region. In contrast, n-type GaAs provides a built-in voltage at the

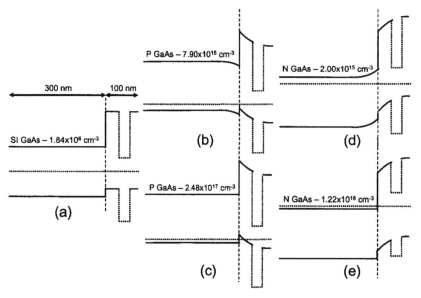

Fig. 4. Band diagrams showing the depletion region at the heterojunction interface for (a) SI, (b) p-type, 7.9e15cm⁻³, (c) p-type 2.48e17 cm⁻³, (d) n-type 2e15 cm⁻³, and (e) n-type 1.22e18 cm⁻³ GaAs substrates.

p-n junction that serves to block electron injection.

This view of the CQD/polymer nanocomposite on semiconductor substrates suggests the following mechanisms enabling IR absorption. Band-bending and carrier depletion at the semiconductor substrate/conducting polymer heterojunction provides electrons that can occupy confined energy levels in the CQD conduction band. In addition, the semiconductor substrate absorbs incident, near-IR light from the FTIR source, photo-generating carriers that can also provide electrons to confined energy levels in the CQD conduction band (especially on p-type GaAs). Holes are blocked from injection into confined CQD states due to Type-II band line-up, and tend to occupy CQD surface states or are transported by MEH-PPV. The localization of electrons in the confined levels enables intraband absorption across a wide IR range.

4. Conclusions

Intraband transitions in three-dimensional, quantum-confined materials (QDs) can significantly increase the operating temperature of IR photodetectors, thereby reducing the cost of IR cameras. Controlling dopant incorporation in InAs/GaAs QDs is important to the optimization of QDIP device performance. By temperature-dependent dark current-voltage measurements in samples using different doping schemes, we have provided experimental evidence for the important contribution of thermally-activated, defect-assisted, sequential resonant tunneling in QDIPs. This dark current mechanism may be related to DX centers in GaAs, which have been observed using polarization-dependent, FTIR spectroscopy. Using MEH-PPV as a matrix for CdSe CQDs enhances quantum confinement and electron localization in the conduction band. As a result, CQDs in the intermediate-confinement regime can display intraband transitions in the mid-IR. In addition, these IR transitions are observed at room temperature, which has important implications for infrared photodetectors. Further, by using semiconductor substrates, and controlling the substrate material, doping type, and doping level, the intraband, IR response can be tuned. Future work will focus on combining InAs/GaAs QDs and CdSe/MEH-PPV nanocomposites on a single GaAs substrate to achieve multi-spectral detection in the IR.

5. Acknowledgments

This work is supported in part by the National Science Foundation CAREER Award and the Air Force Office of Scientific Research, Air Force Research Lab Nano-Initiative. The author thanks her collaborators, Dr. Changhyun Yi at Duke University (MBE growth of InAs/GaAs QDIPs) and Prof. Claire Gmachl at Princeton University (Polarization-dependent FTIR measurements of InAs/GaAs QDIPs). The author also acknowledges Zhiya Zhao and Udoka Uzoka, graduate and undergraduate students, respectively, that have contributed to this work.

References

1. A. D. Stiff-Roberts, S. Chakrabarti, X. Su et al., "Research propels quantum dots forward," Laser Focus World **41**, 103-108 (2005).
2. S. Krishna, "Quantum dots-in-a-well infrared photodetectors," J. Phys. D: Appl. Phys. **38**, 2142 (2005).
3. P. Bhattacharya, X. H. Su, S. Chakrabarti et al., "Characteristics of a tunneling quantum-dot infrared photodetector operating at room temperature," Appl. Phys. Lett. **86**, 191106 (2005).
4. P. Bhattacharya, S. Ghosh, and A. D. Stiff-Roberts, "Quantum dot opto-electronic devices," Annual Review of Materials Research **34**, 1-40 (2004).
5. C. B. Murray, S. Sun, W. Gaschler et al., "Colloidal synthesis of nanocrystals and nanocrystal superlattices," IBM Journal of Research and Devices **45** (1), 47-56 (2001).
6. W. W. Yu, L. Qu, W. Guo et al., "Experimental Determination of the Extinction Coefficient of CdTe, CdSe, and CdS Nanocrystals," Chemical Materials **15**, 2854-2860 (2003).
7. A. P. Alivisatos, "Perspectives on the physical chemistry of semiconductor nanocrystals," Journal of Physical Chemistry **100**, 13226-13239 (1996).
8. D. J. Binks, "Quasi-Bound State Theory of Field-Dependent Photogeneration from Polymer-Embedded Nanoparticles," IEEE Journal of Quantum Electronics **40** (8), 1140-1149 (2004).
9. A.D. Stiff-Roberts, University of Michigan, 2004.

10. E. Calleja and E. Munoz, in *Physics of DX Centers in GaAs Alloys*, edited by J.C. Bourgoin (Sci-Tech Publications, Haus Gafadura, 1990), pp. 73-98.

11. E. Munoz, E. Calleja, I. Izpura et al., "Techniques to minimize DX center deleterious effects in III-V device performance," J. Appl. Phys. **73**, 4988 (1993).

12. M.-H. Du and S.B. Zhang, "DX Center in GaAs and GaSb," Phys. Rev. B **72**, 075210 (2005).

13. L. Bakueva, S. Musikhin, M. A. Hines et al., "Size-tunable infrared (1000-1600nm) electroluminescence from PbS quantum-dot nanocrystals in a semiconducting polymer," Applied Physics Letters **82** (17), 2895-2897 (2003).

14. N. Tessler, V. Medvedev, M. Kazes et al., "Efficient Near-Infrared Polymer Nanocrystal Light-Emitting Diodes," Science **295**, 1506-1508 (2002).

15. C. A. Leatherdale, C. R. Kagan, N. Y. Morgan et al., "Photoconductivity in CdSe quantum dot solids," Physical Review B **62** (4), 2669-2680 (2000).

16. P. Guyot-Sionnest and M. A. Hines, "Intraband transitions in semiconductor nanocrystals," Applied Physics Letters **72**, 686-688 (1998).

17. D. S. Ginger, A. S. Dhoot, C. E. Finlayson et al., Applied Physics Letters **77**, 2816 (2000).

18. M. I. Vasilevskiy, A. G. Rolo, M. V. Artemyev et al., Phys. Stat. Sol. B **224**, 599 (2001).

19. J. M. Pietryga, R. D. Schaller, D. Werder et al., "Pushing the Band Gap Envelope: Mid-Infrared Emitting Colloidal PbSe Quantum Dots," Journal of American Chemical Society Communications **126**, 11752-11753 (2004).

20. S. A. McDonald, P. W. Cyr, L. Levina et al., "Photoconductivity from PbS-nanocrystal/semiconducting polymer composites for solution-processible, quantum-size tunable infrared photodetectors," Applied Physics Letters **85** (11), 2089-2091 (2004).

21. K. R. Choudhury, Y. Sahoo, T. Y. Ohulchanskyy et al., Applied Physics Letters **87**, 073110-073111 (2005).

International Journal of High Speed Electronics and Systems
Vol. 17, No. 1 (2007) 173–176
© World Scientific Publishing Company

BALLISTIC ELECTRON ACCELERATION
NEGATIVE-DIFFERANTIAL-CONDUCTIVITY DEVICES

BARBAROS ASLAN

*School of Electrical and Computer Engineering, Cornell University, 424 Phillips Hall,
Ithaca, NY 14853, USA*
ba58@cornell.edu

LESTER F. EASTMAN[1], WILLIAM J. SCHAFF[1], XIAODONG CHEN[1],
MICHAEL G. SPENCER[1], HO-YOUNG CHA[2], ANGELA DYSON[3], BRIAN K. RIDLEY[3]

[1] School of Electrical and Computer Engineering, Cornell University, Ithaca, NY 14853, USA
[2] General Electric R&D Lab, Schenectady
[3] Electronic Systems Engineering, University of Essex, Colchester, UK

We present the experimental development and characterization of GaN ballistic diodes for THz operation. Fabricated devices have been described and gathered experimental data is discussed. The major problem addressed is the domination of the parasitic resistances which significantly reduce the accelerating electric field across the ballistic region (intrinsic layer).

Keywords: Terahertz; Ballistic Transport; Negative Differential Conductivity; Negative Differential Resistance; GaN Diode

1. Introduction

There is currently considerable interest in compact sources of THz radiation. Applications are numerous: security, medical and environmental sensing. Many novel approaches have been suggested to fill the THz gap, a frequency range that is difficult to cover with conventional solid state devices. However, room temperature operation remains a major challenge.

Possibility of negative-differential-resistance (NDR) is suggested by Ridley et al., showing that electrons in an ultrashort piece of intrinsic GaN (~20nm) can be ballistically accelerated to the Brillouin Zone energy limit of 2.7eV. Energized beyond the inflection point (1.0eV), they reach the negative effective mass states, allowing NDR.[1, 2] Experiments by M. Wraback indicate the presence of such phenomena. [3]

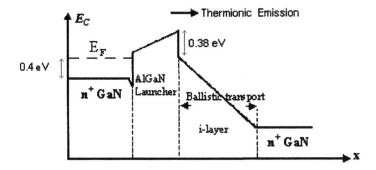

Fig. 1.1 Band diagram for AlGaN/GaN launcher structures. Electrons are launched with 0.38 eV energy.

2. Experiments

The first fabricated devices were simple n^+-i-n^+ structures with 2×10^{19}/cm^3 to 6×10^{19}/cm^3 n^+ doping. It was clear from early theoretical work that these structures would not be optimum since 'hot injection' of electrons was preferred. However these devices provided some very useful information during the development. An AlGaN/GaN launching mechanism was considered next and fabricated to achieve hot injection. The band diagram for this structure is shown in Fig. 1.1

Although the theory prescribes 20nm of i-layer thickness for no collisions, practical limitations allowed us to study 30nm. The diodes on both structures have circular active regions defined by a dry etch process (chlorine based ICP). The diameters were chosen as 3, 5, 10, 15 and 25 microns (Fig. 3.1). After the deposition of ohmic contacts, a standard airbridge process is utilized to connect the diodes to larger cascade probing pads. Parasitic resistances were monitored through on chip TLM patterns. Moreover; planar, no-mesa diagnostic devices of same diameter have also been fabricated in close proximity of the actual diodes in order to track the amount of bias voltage dropping across the parasitics (Fig. 3.2).

3. Results and Discussion

Characterization of these first n^+-i-n^+ structures revealed certain drawbacks regarding the implementation of the theory. A major difficulty in getting enough of the bias voltage across the i-layer has been encountered. This is very important for energizing the electrons negative mass states. Two main issues are: (i) High contact resistances capturing most of the applied bias voltage. (ii) Spreading resistance reducing the field at the center of the i-layer.

The high contact resistances to n^+ regions were unexpected because the same ohmic contact technology used for AlGaN/GaN FET's was employed. Metal stack used for this contact is Ti/Al/Mo/Au which is alloyed at 800C for 30secs. The domination of the parasitic resistances in the device behavior is seen for the continuous wave (C.W.) I-V comparison of an actual device with a planar no-mesa diagnostic version (geometry shown in Fig. 3.2).

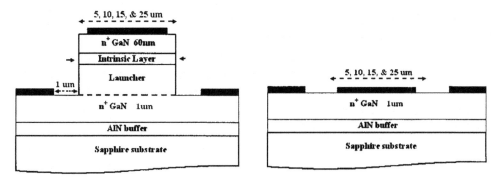

Fig. 3.1 Device Geometry

Fig. 3.2 Diagnostic device geometry

Measurements indicate most of the bias voltage drops across the parasitics instead of the i-layer. Moreover, bias voltages above approximately 3 Volts result in heating at the contact to a high enough temperature to vaporize the connection at the air bridge. In an effort to isolate thermal effects from the genuine I-V characteristics, pulsed and low temperature measurements are also performed. A series of TLM experiments with various metal stacks were performed in search of better contacts. Among these, *non-alloyed* Ti/Al/Mo/Au proved to be the best with a specific contact resistance of 4.92×10^{-8} Ω-cm^2 for 1.26×10^{20}/cm^3 n+ doping, enabling a hundred to one improvement over the previous values. In addition, the geometry of the mesa's produced a significant parasitic spreading resistance. This was due to the large physical gap between the mesa and the cathode. Spreading resistance is responsible for reducing the electric field towards the center of the mesa at the launcher – intrinsic layer interface.

As part of the efforts to solve these problems, a higher doping of 1×10^{20}/cm^3 was incorporated in the design of second generation devices. These utilized AlGaN/GaN barriers as the launching mechanism (see Fig. 1.1) for higher injection energy. Non-alloyed Ti/Al/Mo/Au ohmic contacts are used. I-V curves for these devices are presented in Fig. 3.3. These curves are obtained through successive voltage sweeps of the same 10um diode. As can be seen from the figure, a dc negative differential resistance is observed approximately around 1.9 bias voltage for the first and second sweeps. However, *as-deposited* nature of the aluminum containing metal stack is considered to be responsible for the degradation of the contact resistances due to extensive heating, especially at the top contact. It is not conclusive how much of a role the unstable, high resistance anode ohmic contact plays in the observation of the negative differential resistance.

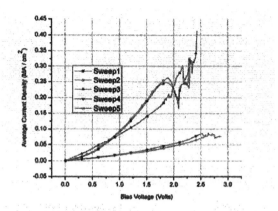

Fig. 3.3 Successive I-V sweeps of a 10um diameter device on a %20 AlGaN/GaN launcher design.

4. Conclusion

Details of the development of the GaN ballistic diode have been presented. Experimental results indicate the high contact resistance's on highly doped GaN are a major problem. TLM experiments showed significant reduction of contact resistances through the use of non-alloyed metal stacks. However, the stability of such contacts under high current density operation still needs to be addressed.

5. Acknowledgement

This work has been funded by ONR (Project Monitor Dr. Paul Maki). We would also like to express our gratitude to Dr. Colin Wood (former Project Monitor) for his support at the earlier stages of the research.

6. References

1. B. K. Ridley, W. J. Schaff & L.F. Eastman, "Theory of the GaN crystal diode: Negative mass negative differential resistance", *J. Appl. Phys.*, vol.97, pp. 094503-094509, (2005)
2. C. Bulutay, B. K. Ridley, and N. A. Zakhleniuk, "Electron momentum and energy relaxation rates in GaN and AlN in the high-field transport regime", Phys. Rev. B 68, 115205 (2003). [ISI]
3. M. Wraback, H. Shen, J. C. Carrano, T. Li, J. C. Campbell, M. J. Schurman and I.T. Ferguson, "Time resolved electro absorption measurement of the transient electron velocity overshoot in GaN", *Appl. Phys. Lett.* 76, pp 1155 (2000)

Section V.
Photoluminescence and Photocapacitance

International Journal of High Speed Electronics and Systems
Vol. 17, No. 1 (2007) 179–188
© World Scientific Publishing Company

World Scientific
www.worldscientific.com

UNDERSTANDING ULTRAVIOLET EMITTER PERFORMANCE USING INTENSITY DEPENDENT TIME-RESOLVED PHOTOLUMINESCENCE

MICHAEL WRABACK, GREGORY A. GARRETT, ANAND V. SAMPATH, PAUL H. SHEN

U.S. Army Research Laboratory, Sensors and Electron Devices Directorate,
2800 Powder Mill Road, Adelphi, MD
mwraback@arl.army.mil

Time-resolved photoluminescence studies of nitride semiconductors and ultraviolet light emitters comprised of these materials are performed as a function of pump intensity as a means of understanding and evaluating device performance. Comparison of time-resolved photoluminescence (TRPL) on UV LED wafers prior to fabrication with subsequent device testing indicate that the best performance is attained from active regions that exhibit both reduced nonradiative recombination due to saturation of traps associated with point and extended defects and concomitant lowering of radiative lifetime with increasing carrier density. Similar behavior is observed in optically pumped UV lasers. Temperature and intensity dependent TRPL measurements on a new material, AlGaN containing nanoscale compositional inhomogeneities (NCI), show that it inherently combines inhibition of nonradiative recombination with reduction of radiative lifetime, providing a potentially higher efficiency UV emitter active region.

Keywords: GaN; AlGaN; femtosecond; time-resolved photoluminescence; luminescence downconversion; carrier localization; carrier lifetime; compositional fluctuations; radiative lifetime; radiative efficiency; nonradiative recombination.

1. Introduction

Ultraviolet light sources based on III-nitride semiconductors may be instrumental in the development of compact, lightweight, low-cost, low-power-consumption systems addressing biological analysis and water/surface treatment, as well as in lighting and high density data storage and retrieval applications. Typical III-Nitride semiconductor-based UV light-emitting diodes (LEDs) have wall plug efficiencies and lifetimes far less than those of commercially available blue LEDs[1]. These limitations are mainly due to a large density of defects caused by heteroepitaxial growth of these III-Nitride- based devices on lattice mismatched substrates, which leads to enhanced nonradiative recombination that reduces their radiative efficiency while increasing debilitating heating. In this paper, we study the correlation of material properties with device performance in state-of-the-art III-Nitride ultraviolet light emitters and detectors using intensity dependent time-resolved photoluminescence (TRPL) on device active regions and full device structures prior to fabrication, with metrics defined as potential performance indicators prior to device processing. We also discuss the potential of nanoscale compositional inhomogeneities (NCI) in AlGaN to improve the quantum efficiency and overall performance of UV light emitters.

2. Experimental

Time-resolved photoluminescence measurements were performed using fs amplified laser pulses continuously tunable between 225 nm and 375 nm. This broad tunability enables one to utilize an excitation pulse that passes through the transparent substrate and n-type current injection layers to excite carriers directly in the active region of an LED structure. The TRPL results can therefore be correlated with the LED output powers and external quantum efficiencies directly. This type of characterization enables one to chart and understand materials growth improvement prior to full device fabrication, while allowing one to separate active region quality from that of current injection layers. Time-resolved data is obtained using two techniques: (1) gated downconversion, with temporal resolution of ~ 300 fs, time scale of ~ 1 ns, and a dynamic range enabling ~ 2 orders of magnitude variation in photoexcited carrier density; and (2) time-correlated single photon counting, with temporal resolution of ~ 25 ps, time scale of up to 4000 ns, and a dynamic range enabling up to 5 orders of magnitude variation in photoexcited carrier density. Complementary time-integrated PL measurements are performed using a continuous wave (CW) 244 nm frequency-doubled argon ion laser. Both time-resolved and time-integrated PL data can be taken in the 10K to 300K temperature range.

3. Time-Resolved Photoluminescence Metrics for UV LED and Laser Performance

Figure 1 shows pump intensity dependent TRPL data obtained using gated downconversion from a representative InAlGaN multiple quantum well (MQW) active region on a GaN template designed for 325 nm emission that illustrates several of the issues faced in UV emitter development. As the areal carrier density is increased from ~5×10^{11} cm^{-2} to ~5×10^{12} cm^{-2}, the PL decay time grows from 351 ps to 544 ps, with a much weaker increase when the intensity is raised by an additional factor of 3. The intensity dependence of the decay times is accompanied by a superlinear increase in the t=0 PL signal I_o, which grows by a factor of ~ 14 when the pump intensity is raised by an order of magnitude, and by a factor of 3.7 when the intensity is increased by an additional 3 times. This sample exhibits a longer PL lifetime than expected for growth on a GaN template with dislocation density in the mid-10^8cm^{-2} range[2]. Since carrier lifetime is strongly influenced by variations in point defect density for a given dislocation density[3], this result suggests that small amounts of indium may improve material quality[4] through reduction in point defects. The behavior of the PL decays and I_o with intensity imply that the large polarization field in these c-plane MQWs[5,6], which induces both spatial separation of the electron and hole wave functions and enhanced wave function penetration in the barriers[7,8], is more effectively screened at higher excitation density[9], increasing the wave function overlap, which decreases the radiative lifetime, while decreasing the amount of wave function in the barrier and saturating nonradiative centers in the MQW. This interpretation is supported by the observation that I_o, which scales with the inverse of the radiative lifetime τ_{rad}, increases superlinearly with the pump intensity, indicating that the radiative lifetime decreases with increasing pump intensity due to better wave function overlap in the wells. Calculations show that wave function penetration into the barriers is more sensitive than electron-hole wave function overlap in the wells to variations in field screening by photogenerated carriers. Therefore, the intensity dependence of the subsequent carrier dynamics is governed by the removal of carriers from the wells through barriers/interface states and saturation of nonradiative centers in the MQW, with the increase of PL lifetime with increasing intensity in the *c*-plane MQWs attributed primarily to an increase in nonradiative lifetime in conjunction with greater field screening.

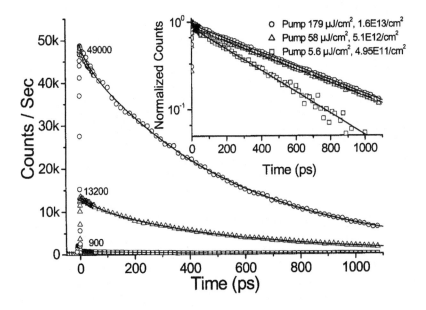

Fig. 1.   Intensity dependent TRPL decays for an InAlGaN MQW designed for emission at 325 nm on a GaN template.

As material quality and associated device performance improve, the pump intensity dependent TRPL data exhibits a different behavior. Figure 2 shows TRPL data obtained using time-correlated single photon counting from the active region of a 280 nm LED device structure prior to fabrication. As the pump intensity increases from 346 nJ/cm^2 to 0.1 mJ/cm^2, the TRPL decay curve remains about the same, exhibiting a complex TRPL decay that is significantly longer than that for the MQW described above, approximated by a dominant 927 ps decay and a weaker 347 ps component. This behavior suggests that the density of nonradiative centers is low enough that they are strongly saturated even at low pump intensity, leading to longer PL lifetimes. At the highest pump intensity (~ 1mJ/cm^2) the decay becomes noticeably faster.   This decrease in photoluminescence decay time may signify the onset of nonlinear radiative recombination corresponding to a reduced radiative lifetime at high carrier density, characteristic of higher quality samples with less defects contributing to nonradiative recombination. 280 nm LEDs fabricated from these wafer batches exhibited ~0.6 -1 mW of output power at 20 mA continuous drive current. These results indicate that TRPL can be an effective diagnostic of device quality prior to fabrication.

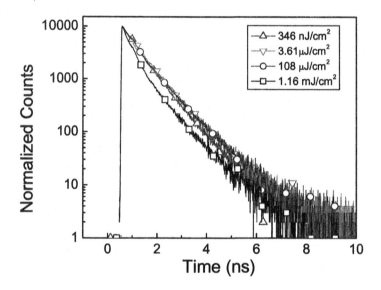

Fig. 2. Intensity dependent TRPL from a 280 nm LED wafer prior to processing.

Figure 3 shows intensity dependent TRPL and time-integrated pulsed PL data from a 330 nm laser structure that further supports the interpretation of Figs. 1 and 2 while combining aspects of both. As the pump intensity is increased by more than four orders of magnitude, the time-integrated pulsed PL exhibits superlinear behavior. The TRPL data show that for lower pump intensities increasing the fluence from ~10 nJ/cm^2 to ~ 1 µJ/cm^2 leads to an increase of the dominant PL lifetime from 635 ps to 1.12 ns, implying that the superlinear increase in pulsed PL is due to saturation of nonradiative centers. The complex decay embodied in the long tail for the lowest fluence may be indicative of a regime in which the nonradiative centers are saturated, but the carrier density does not remain high enough that the internal fields in the MQW are screened, thus leading to a longer radiative lifetime due to the reduction in wave function overlap. The lack of this PL decay tail at the higher fluences is consistent with the more effective screening of the MQW internal electric fields. As the pump intensity is raised by another order of magnitude, the PL decay remains relatively unchanged, as expected for strong field screening and saturation of nonradiative sites, analogous to the 280 nm LED results described earlier. Finally, at the highest fluence, $\sim 142~\mu J/cm^2$, the superlinear increase in the time-integrated pulsed PL correlates with a dramatic drop in PL lifetime to ~ 464 ps, thus indicating the onset of nonlinear radiative recombination associated with a density dependent reduction in radiative lifetime.

4. Enhanced Luminescence Efficiency Through Nanoscale Compositional Inhomogeneities in AlGaN

The TRPL results described above indicate that an approach that would combine

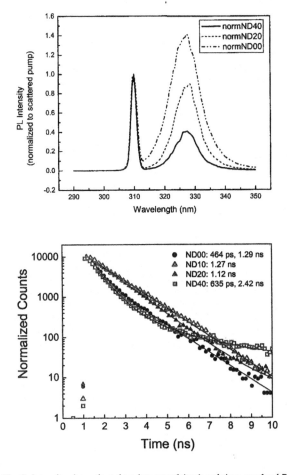

Fig. 3. Intensity-dependent time-integrated (top) and time-resolved PL (bottom) from a 330 nm laser structure.

suppression of nonradiative recombination with shorter radiative lifetime might significantly improve the radiative efficiency in UV LED active regions. Recently[10], we have reported on the development of AlGaN films deposited by plasma assisted molecular beam epitaxy (PA-MBE) that possess enhanced internal quantum efficiency due to the presence of nanometer scale compositional inhomogeneities (NCI-AlGaN) within a wider bandgap matrix that inhibit nonradiative recombination through the large defect densities in these materials. These NCI-AlGaN films exhibit intense room temperature photoluminescence (PL) that is ~ 1000x stronger than, and greater than 250 meV red-shifted with respect to, the normally expected band edge emission. Further insight into this phenomenon can be gleaned from examination of the optical properties of these NCI AlGaN films as a function of red-shift of the dominant PL emission from the bandgap of the surrounding matrix. Figure 4 shows that the room temperature (RT) PL under (CW) excitation increases for larger red-shift under controlled growth conditions

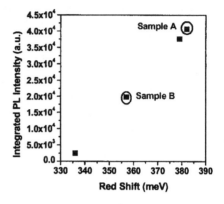

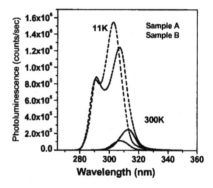

Fig. 4. Photoluminescence (PL) intensity dependence on Stokes shift for NCI AlGaN samples (top); Room and low temperature PL spectra for two samples in the series (bottom).

in thin NCI-AlGaN films on sapphire. However, comparison of two samples in the series shows that sample A, with the larger shift and brighter RT PL, has dimmer PL at low temperature (15K) than sample B. Femtosecond time-resolved PL (TRPL) using gated down-conversion provides a possible explanation of this behavior, since the dynamics of carrier transfer from the band edge matrix to the NCI region can be seen. The fact that sample A has NCI emission twice as bright as sample B under CW excitation is reflected in the longer RT lifetime in Sample A, ~ 459 ps verses 313 ps (Fig. 5). Sample B also has a faster decay in the band edge matrix PL that suggests a more rapid transfer of carriers to a larger density of NCI states in this film than sample A. This explanation is corroborated by the observation that the peak signal for the NCI emission near time-zero for sample A becomes smaller than for sample B with increasing pump intensity due to saturation of the smaller density of NCI states in sample A, and is further supported by the larger PL intensity in sample B at temperatures low enough that nonradiative recombination paths

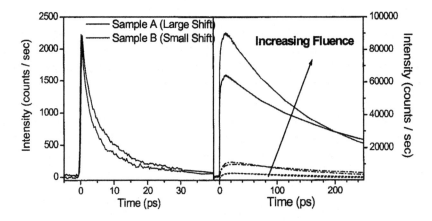

Fig.5. Bandedge luminescence at high pump intensity (left) and pump intensity dependent NCI luminescence (right) as a function of time delay for the two samples.

are frozen out. The shorter RT lifetime at densities high enough to saturate nonradiative centers within the NCI for sample B suggests that the individual NCI regions in this sample may be smaller, with a larger relative density of interface states that may mediate nonradiative recombination through unsaturated defect states in the wider bandgap matrix. However, while the smaller shift of the NCI emission peak in moving from 15K to RT in sample B relative to sample A (64 meV vs. 75 meV) supports the notion that the NCI in sample B may be smaller, with concomitant reduced electron-phonon interaction, the shifts in both samples are too large to invoke quantum confinement[11]. Rather, the

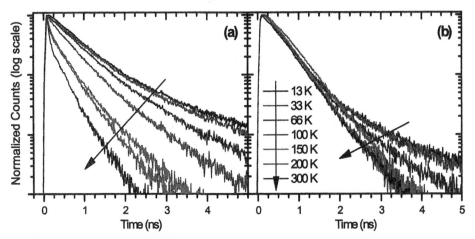

Fig. 6. Time-resolved PL as a function of temperature for (a) bandedge bulk and (b) localized NCI material emission. A pump excitation fluence of 2.8 $\mu J/cm^2$ (est. carrier density of $4\times10^{17}cm^{-3}$) at 285 nm was used.

smaller nonradiative lifetime (333 ps vs. 546 ps) and longer radiative lifetime (5.2 ns vs. 2.9 ns) in sample B relative to sample A under low excitation conditions at room temperature imply that one of the main reasons for the shorter PL lifetime and weaker time-integrated PL in this sample may be the lower carrier density in the NCI associated with the relatively larger density of NCI regions.

An even smaller redshift of the NCI emission might be associated with a relative blue shift due to quantum confinement. Figure 6 shows TRPL on a sample with a redshift of 300 meV, smaller than that in Sample B, but which is brighter than Sample A, most likely due to a thick, possibly lower defect density, base layer on which it was grown. As temperature is decreased from 300 K to 10 K, PL emission from the band edge arises from the noise to about 20% that of the NCI red-shifted peak, and its lifetime increases from a system response limit of 25 ps to over 600 ps. This behavior is related to the freeze out of nonradiative sites and localization of carriers in the bulk, wider bandgap matrix[12,13]. For the NCI peak, an approximately constant lifetime of 450 ps is observed as a function of temperature and an internal quantum efficiency of ~28% is estimated from the ratio of the low- and room-temperature PL intensities. Figure 7 shows the calculated radiative lifetimes for the NCI and bandedge emissions. Since it is difficult to determine the intensity of the bandedge emission corresponding to the wider bandgap matrix at temperatures above 150K, the radiative lifetime for bandedge emission is only an estimated lower bound. On the other hand, the radiative lifetime for NCI represents an upper bound due to a continuous transfer of carriers from the matrix to the NCI region. The radiative lifetime for bandedge emission is consistent with measured values for GaN and with the theoretical value at equivalently doped GaN[14,15]. The radiative lifetime for the red-shifted emission from the NCI is almost independent of temperature, reaching a value of ~ 2 ns at room temperature. One reason for this may be the large amount of carriers transferred from the matrix to the NCI. Theoretical calculation indicates the temperature dependence of the radiative lifetime is significantly reduced when the carrier density is high[14]. On the other hand, it is well known that the radiative lifetime of a zero dimensional exciton is nearly constant over temperature. This suggests a lower dimensionality of the NCI region. This idea is supported by a smaller measured shift in the NCI PL peak, from room to low temperature, of 48 versus 75 meV for the 300 versus 400 meV red-shift samples due to a reduction of the electron-phonon interaction in the smaller structures.

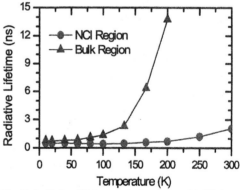

Fig. 7. Radiative lifetimes calculated from PL lifetimes and estimated internal quantum efficiencies.

Further insight can be obtained from the room temperature, intensity dependent TRPL data in Fig. 8 from a double heterostructure (DH) LED structure on an AlGaN template with the NCI AlGaN as the active region. In this case, the pump pulse is tuned to 265 nm to include photogeneration of carriers in the barrier region of the structure. The PL lifetime at low fluence (< 20 μJ/cm^2) increases by nearly a factor of two in the DH LED structure relative to similar layers on sapphire, with a half-life of ~ 500 ps,

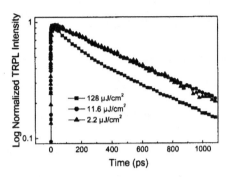

Fig. 8.   Intensity dependent room temperature TRPL from a DH LED structure taken at the 330 nm peak corresponding to NCI emission.

and a dominant decay time in excess of 700 ps. Although this sample has dislocation density in the mid 10^{10}cm^{-2} range, the decay times are comparable to those measured in high quality GaN grown atop HVPE GaN templates with dislocation densities in the mid 10^7cm^{-2} range[2,3], suggesting that the further suppression of nonradiative recombination processes within and at the boundaries of the NCI regions also contributes to the increased internal quantum efficiency in the DH LED structure. While the PL decay shows no intensity dependence at low densities, it exhibits a dramatic decrease in the half-life at the highest fluence (128 μJ/cm^2) to 292 ps that is associated with the onset of a fast decay component. This drop in the PL decay time, as seen above in the high quality 280 nm LED active regions, may be indicative of a decrease in the radiative lifetime at high carrier densities due to nonlinear radiative recombination in the localized regions, which may be driven by enhanced carrier concentrations associated with the transfer of carriers from the AlGaN matrix to the smaller localized regions and the suppression of nonradiative processes in these regions. Thus, the combination of saturation of nonradiative recombination sites through carrier transfer from the matrix and concentration in the NCI and concomitant lower radiative lifetimes of the NCI relative to the matrix, associated with the higher concentration in and possibly lower dimensionality of the NCI, provides an attractive option for development of high efficiency UV LEDs.

5. Conclusion

We have used pump intensity dependent time-resolved and time-integrated photoluminescence as a means of defining performance metrics for state-of-the-art UV light sources. Comparison of TRPL on UV LED wafers prior to fabrication with subsequent device testing and on UV lasers indicate that the best performance is attained from active regions that exhibit both reduced nonradiative recombination due to saturation of traps associated with point and extended defects and concomitant lowering of radiative lifetime with increasing carrier density. Temperature and intensity dependent TRPL measurements on a new material, AlGaN containing nanoscale compositional inhomogeneities (NCI), show that it inherently combines inhibition of nonradiative recombination with reduction of radiative lifetime, providing a potentially higher efficiency UV emitter active region.

Acknowledgements

The authors would like to thank N.M. Johnson of PARC for the InAlGaN MQW and laser samples and helpful discussions, T. Katona of SET, Inc. for 280 nm LED samples and helpful discussions, and V. Dmitriev of TDI, Inc. for the AlGaN template used for the NCI AlGaN DH LED structure and helpful discussions.

References

1. J. Zhang, X. Hu, A. Lunev, J. Deng, Y. Bilenko, T.M. Katona, M.S. Shur, R. Gaska, and M.A. Khan, Japan. J. Appl. Phys. Part 1 **44**, 7250 (2005), and references therein.
2. H.K. Kwon, C.J. Eiting, D.J.H. Lambert, M.M. Wong, R.D. Dupuis, Z. Liliental-Weber, and M. Benamara, Appl. Phys. Lett. **77**, 2503, (2000).
3. A.V. Sampath, G.A. Garrett, C.J. Collins, P. Boyd, J. Choe, P.G. Newman, H. Shen, M. Wraback, R.J. Molnar, and J. Caissie, J. Vac. Sci. Technol. **B22**, 1487 (2004).
4. G. Tamulaitis, K. Kazlauskas, S. Jursenas, A. Zukauskas, M.A. Khan, J.W. Yang, J. Zhang, G. Simin, M.S. Shur, and R. Gaska, Appl. Phys. Lett. **77**, 2136 (2000).
5. F. Bernardini, V. Fiorentini, and D. Vanderbilt, Phys. Rev. **B56**, R10024 (1997).
6. J.S. Im, H. Kollmer, J. Off, A. Sohmer, F. Scholz, and A. Hangleiter, Phys. Rev. **B57**, R9435 (1998).
7. N. Grandjean, B. Damilano, S. Dalmasso, M. Leroux, M. Laügt, and J. Massies, J. Appl. Phys. **86**, 3714 (1999).
8. Y.D. Jho, J. S. Yahng, E. Oh, and D. S. Kim, Phys. Rev. **B66**, 035334 (2002).
9. P. Lefebvre, S. Kalliakos, T. Bretagnon, P. Valvin, T. Taliercio, B. Gil, N. Grandjean, and J. Massies, Phys. Rev. **B69**, 035307 (2004).
10. C. J. Collins, A.V. Sampath, G.A. Garrett, W.L. Sarney, H. Shen, M. Wraback, A.Yu. Nikiforov, G.S. Cargill, III, and V. Dierolf, Appl. Phys. Lett. **86**, 31916 (2005).
11. J. Brown, C. Elsass, C. Poblenz, P.M. Petroff, and J.S. Speck, Phys. Stat. Sol. (b) **228**, 199 (2001).
12. Y.-H. Cho, G.H. Gainer, J.B. Lam, and J.J. Song, Phys. Rev. **B61**, 7203 (2000).
13. H.S. Kim, R.A. Mair, J. Li, J.Y. Lin, and H.X. Jiang, Appl. Phys. Lett. **76**, 1252 (2000).
14. J. S. Im, A. Moritz, F. Steuber, V. Harle, F. Scholz, and A. Hangleiter, Appl. Phys. Lett. **70**, 631 (1997).
15. O. Brandt, J. Ringling, K.H. Ploog, H.-J. Wunsche, and F. Henneberger, Phys. Rev. **B58**, R15977 (1998).

International Journal of High Speed Electronics and Systems
Vol. 17, No. 1 (2007) 189–192
© World Scientific Publishing Company

PHOTOCAPACITANCE OF SELECTIVELY DOPED AlGaAs/GaAs HETEROSTRUCTURES CONTAINING DEEP TRAPS

NIKOLAI B. GOREV, INNA F. KODZHESPIROVA, EVGENY N. PRIVALOV

Department for Functional Elements of Control Systems, Institute of Technical Mechanics, National Academy of Sciences of Ukraine, 15 Leshko-Popel' St., Dnepropetrovsk, 49005, Ukraine
gorev57@mail.ru

NINA KHUCHUA, LEVAN KHVEDELIDZE

Research and Production Complex "Electron Technology", Tbilisi State University, 13 Chavchavadze Ave., Tbilisi, 380079, Georgia
nt@gol.ge

MICHAEL S. SHUR

Center for Broadband Data Transport, Room 9017, CII, Rensselaer Polytechnic Institute, 110 8-th Street, Troy, NY 112180-3590, USA
shurm@rpi.edu

The results of calculations of the low-frequency and the high-frequency barrier capacitance of selectively doped AlGaAs/GaAs heterostructures containing deep traps in the AlGaAs layer are presented. The calculations are done for the samples in the dark and under extrinsic illumination. It is shown that the high-frequency photocapacitance of these structures exhibits a positive peak, and the low-frequency photocapacitance has a positive peak followed by a negative valley. The underlying physical mechanisms are discussed.

Keywords: Heterostructure; barrier capacitance; photocapacitance; deep traps.

1. Introduction

High electron mobility transistors (HEMTs) are widely used in modern microelectronics. Of considerable interest, in terms of both characterization purposes and the development of optically controlled devices, is the response of the barrier capacitance of a HEMT heterostructure to optical illumination, i.e., its photocapacitance. It should be noted at this point that HEMT heterostructures have a doping step at the heterointerface and usually contain deep traps (thus AlGaAs/GaAs HEMT heterostructures usually contain deep donors (*DX* centers) in the AlGaAs layer). As we showed earlier by the example of metal–semiconductor field-effect transistor (MESFET) structures[1], steep dopant gradients in combination with the presence of deep traps may radically alter the behavior of photocapacitance in comparison with graded dopant profile structures, for which the majority of existing photocapacitance techniques have been developed. The goal of this paper is to study the photocapacitance of typical selectively doped AlGaAs/GaAs heterostructures with deep donor traps in the AlGaAs layer.

2. Calculation of Barrier Capacitance and Photocapacitance

The band diagram of a typical selectively doped AlGaAs/GaAs heterostructure (metal – n-$Al_xGa_{1-x}As$ – p-GaAs) is shown in Fig. 1. The $Al_xGa_{1-x}As$ layer of thickness d contains shallow donors with concentration N_D and ionization energy W_D and deep donor-like electron traps with concentration N_t and ionization energy W_t, while the GaAs layer contains shallow acceptors with concentration N_A. Let a reverse bias, V_{rev}, be applied to the Schottky contact.

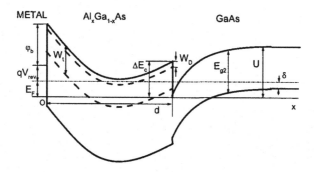

Fig. 1. Band diagram of a selectively doped $Al_xGa_{1-x}As$/GaAs heterostructure.

The low-frequency (LF) barrier capacitance C_{LF} and the high-frequency (HF) barrier capacitance C_{HF} are related to the electric field E_b at the metal/$Al_xGa_{1-x}As$ interface as follows: $C_{LF} = \varepsilon_1 S_b \, dE_b/dV_{rev}$ and $C_{HF} = \varepsilon_1 S_b \, \delta E_b/\delta V_{rev}$, $\delta n_t = 0$. Here, ε_1 is the $Al_xGa_{1-x}As$ permittivity (hereinafter the subscripts 1 and 2 will refer to the $Al_xGa_{1-x}As$ and the GaAs layer, respectively), δE_b and δn_t are the variations of E_b and of the trapped carrier density, respectively, with the variation δV_{rev} of the reverse bias, and S_b is the barrier contact area.

Poisson's equation for the $Al_xGa_{1-x}As$ layer and the initial conditions for this equation at the heterointerface are

$$\frac{d^2\varphi}{dx^2} = \frac{q}{\varepsilon_1}\left(N_D^+ + N_t^+ - n\right), \qquad \varphi(d) = \Delta E_c/q, \qquad \varphi'(d) = -E_{j1}. \tag{1}$$

Here, q is the electron charge, $\varphi(x)$ is the electrostatic potential, N_D^+ and N_t^+ are the concentrations of ionized shallow and deep donors, respectively, n is free carrier density, ΔE_c is the conduction band discontinuity and E_{j1} is the electric field in the heterojunction plane on the $Al_xGa_{1-x}As$ side.

The GaAs layer is described using the model proposed by Kal'fa[2]. This model accounts for the entire energy spectrum of the two-dimensional electron gas (2DEG) in the approximation of a triangular potential well. It gives:

$$U = \frac{q^2}{2\varepsilon_2} \frac{1}{N_A} \left(\frac{\varepsilon_2 |E_{j2}|}{q} - \sum_i n_{si} \right)^2 + \frac{2q}{3\varepsilon_2 |E_{j2}|} \sum_i n_{si} E_i,$$

$$n_{si} = \frac{m^* kT}{\pi \hbar^2} \ln \left(1 + \exp \frac{E_F - E_i}{kT} \right), \qquad E_i = \left(\frac{\hbar^2}{2m^*} \right)^{1/3} \left[\frac{3}{2} \pi q |E_{j2}| \left(i + \frac{3}{4} \right) \right]^{2/3}. \qquad (2)$$

Here, E_F is the Fermi energy, U is the GaAs conduction band bending (see Fig. 1), ε_2 is the GaAs permittivity, E_i ($i = 0, 1, ...$) is the eigen energy that corresponds to the bottom of the ith two-dimensional subband, n_{si} is the free carrier sheet density in the ith subband, E_{j2} is the electric field in the heterojunction plane on the GaAs side and m^* is the effective mass of an electron in GaAs.

To Eqs. (1)–(2) must be added the relationships that follow from Fig. 1 and the electric flux continuity at the heterointerface

$$q\varphi(0) = E_F + \varphi_b + qV_{rev}, \qquad E_F = U - E_{g2} + \delta, \qquad \varepsilon_1 E_{j1} = \varepsilon_2 E_{j2}. \qquad (3)$$

Here, φ_b is the Schottky barrier height, and the meaning of the other quantities is clear from Fig. 1.

Equations (1)–(3) can be solved numerically to yield the electric field E_b at the metal–semiconductor interface and thus C_{LF} and C_{HF}.

The LF and HF barrier capacitances of a typical $Al_xGa_{1-x}As/GaAs$ HEMT heterostructure ($x = 0.3$; $d = 0.03$ μm; $N_D = 2 \times 10^{18}$ cm^{-3}; $W_D = 0.01$ eV; $N_t = 6 \times 10^{17}$ cm^{-3}; $W_D = 0.4$ eV; $N_A = 10^{14}$ cm^{-3}; $\varphi_b = 0.8$ eV and $T = 300$ K) calculated using the Fermi–Dirac and the Maxwell–Boltzmann statistics for the $Al_xGa_{1-x}As$ layer are shown in Fig. 2(a) as solid and dashed curves, respectively. As illustrated, the barrier capacitance can be calculated using the Maxwell–Boltzmann statistics from a relatively low reverse bias on (to provide efficient voltage control of the 2DEG, the $Al_xGa_{1-x}As$ layer in HEMT heterostructures is usually so thin that the depletion regions of the Schottky barrier and of the heterojunction merge to a large degree even at zero barrier bias). For the Maxwell–Boltzmann statistics, extrinsic illumination is accounted for by replacing the characteristic concentration $n_1 = 1/2 N_{c1} \exp(-W_t / kT)$ with $n_{1l} = n_1 + \Delta n_1$ where the quantity Δn_1 depends on the illumination intensity and the trap parameters[1].

3. Results and Discussion

The LF and HF barrier capacitances of the $Al_{0.3}Ga_{0.7}As/GaAs$ heterostructure under extrinsic illumination calculated at $n_{1l}/n_1 = 7$ are shown in Fig. 2(a) as dotted curves, and the photocapacitance, i.e., the difference between the capacitance under illumination and the dark capacitance calculated using the Maxwell–Boltzmann statistics, is shown in Fig. 2(b). As illustrated, the dependence of the HF photocapacitance on the reverse bias has the form of a positive peak, and that of the LF photocapacitance has the form of a positive peak followed by a negative valley.

The peaks are due to the photoionization of deep traps in the $Al_xGa_{1-x}As$ layer providing an additional inflow of free carries into the quantum well. Because of this, under illumination the steeply decreasing region of the *C–V* curve (associated with quantum well emptying[3]) starts at a higher reverse bias than in the dark. As the reverse bias is increased, the $Al_xGa_{1-x}As$ layer becomes nearly depleted of free carriers, and the quantum well becomes nearly emptied. From this point on, the predominant contribution to the LF barrier capacitance is caused by the variation of the trapped charge at the edge of the Schottky barrier depletion region in the $Al_xGa_{1-x}As$ layer with reverse bias. Under illumination, the density of this trapped charge decreases, and its decrease approaches zero in two limiting cases: (i) when the free carrier density at the edge of the Schottky barrier depletion region is so high that the deep traps situated there are all occupied even under illumination; (ii) when this free carrier density is so low that the traps are all empty. As a consequence, the LF photocapacitance at high reverse biases is negative and passes through a minimum.

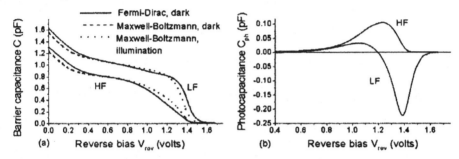

Fig. 2. Calculated LF and HF capacitances and photocapacitances of an $Al_{0.3}Ga_{0.7}As/GaAs$ heterostructure.

4. Conclusions

We have shown that the high-frequency extrinsic photocapacitance of $Al_xGaAs_{1-x}/GaAs$ HEMT heterostructures as a function of reverse bias exhibits a positive peak while the low-frequency photocapacitance exhibits a positive peak followed by a negative valley. These features of the photocapacitance are linked ultimately to the presence of a built-in space charge region stemming from a steep impurity gradient, which is also true for the photocapacitance of GaAs MESFET structures, though their physics is quite different.

References

1. N. B. Gorev, I. F. Kodzhespirova, E. N. Privalov, N. Khuchua, L. Khvedelidze and M. S. Shur, Photocapacitance of GaAs thin-film epitaxial structures, *Solid-State Electron.*, **49**(3), 343–349 (2005).
2. A. A. Kal'fa, Two-dimensional electron gas in a selectively doped metal–$Al_xGa_{1-x}As$–GaAs structure, *Sov. Phys. Semicond.*, **20**(3), 294–297 (1986).
3. V. A. Aleshkin, E. V. Demidov, B. N. Zvonkov, A. V. Murel' and Yu. A. Romanov, Study of quantum wells by the *C–V* method, *Sov. Phys. Semicond.*, **25**(6), 631–636 (1991).

AUTHOR INDEX